Environmental Attitudes T]

Our attitudes to our environment are widely and often acrimoniously discussed, commonly misunderstood, and will shape our future. We cannot assume that we behave as newly minted beings in a pristine garden nor as pre-programmed automata incapable of rational responsibility.

Professor Berry has been involved with many national and international decision-making bodies that have influenced our environmental attitudes. He is therefore well-placed to describe what has moulded our present attitudes towards the environment. This book presents data and concepts from a range of disciplines – genetics, anthropology, sociology, history and theology – to help us understand past responses and how these affect our future. With a historical overview and a discussion of the current situation, this book informs decisions that will have profound impacts on all of us both today and in the years to come.

R.J. BERRY DSC, FRSE, is Professor Emeritus of Genetics, University College London. He has been president of: the Linnean Society, the British Ecological Society, the European Ecological Federation, the Mammal Society and Christians in Science. Professor Berry's research focuses on genetics and ecology. He has received both a Marsh Award for Ecology and a UK Templeton Award.

Environmental Attitudes Through Time

R.J. BERRY

University College London

CAMBRIDGE
UNIVERSITY PRESS

CAMBRIDGE
UNIVERSITY PRESS

University Printing House, Cambridge CB2 8BS, United Kingdom

One Liberty Plaza, 20th Floor, New York, NY 10006, USA

477 Williamstown Road, Port Melbourne, VIC 3207, Australia

314–321, 3rd Floor, Plot 3, Splendor Forum, Jasola District Centre,
New Delhi – 110025, India

79 Anson Road, #06-04/06, Singapore 079906

Cambridge University Press is part of the University of Cambridge.

It furthers the University's mission by disseminating knowledge in the pursuit of
education, learning, and research at the highest international levels of excellence.

www.cambridge.org
Information on this title: www.cambridge.org/9781107062320
DOI: 10.1017/9781107449879

© Cambridge University Press 2018

First published 2018

Printed in the United Kingdom by TJ International Ltd. Padstow Cornwall

A catalogue record for this publication is available from the British Library.

ISBN 978-1-107-06232-0 Hardback
ISBN 978-1-107-67948-1 Paperback

Contents

Preface

This book is not a straightforward account of how our attitudes to the environment have changed through history, although history comes into it because the changes described herein have necessarily occurred in time. Neither is it yet another lament about the environmental damage that is accumulating round us, despite Panglossian deniers attempting to persuade us that all is well. Rather, it is a review of our response to the factors which have shaped us since the time when we became fully human five million years or so ago. Perhaps the best way to regard it is as an annotated chronicle – perhaps idiosyncratic but I hope reasonably objective – of the challenges from the environment as seen by one person (me) embedded within a clamour of fellow-travellers and dissenters. In some ways it is like John Bunyan's *Pilgrim's Progress* in describing landmarks on a journey, but that parallel fails because this particular journey does not have an obvious goal, unlike Bunyan's pilgrim. Nor is it a journey where it is better to travel than to arrive, because of the impending avalanche dangers that line it. Furthermore, it has proved impossible to follow single-mindedly a definitive route, and it has often been necessary to explore (as it were) the scenery around the path. It would be good to think that this would lead to a clear conclusion, but I am only too aware that the end-result is messy and has many loose ends. I have no doubt that I will be condemned for omissions which some will regard as significant or even essential. There are many other topics I might have included, but this is a travelogue, not a textbook. The best I can do is to hope that there will be general agreement about the overall direction of travel and the adjustments we have had to make in the past, and that this will help towards an awareness of the challenges of the present.

My involvement with the issues described in this book
date back nearly half a century when growing awareness of the
environment, particularly its newly recognized importance
following the European Environmental Year in 1970 and the
imminent United Nations Conference on the Human Environment
(1972), prompted a publisher to invite me to write a short book
about environmental ethics – a task which I interpreted somewhat
liberally as going well beyond a branch of academic philosophy
(which it is) but more radically as exploring the no-man's land
where philosophy, psychology, politics, history and management
economics converge. I found myself pitch-forked into a world
where science met ethics and both met action (and inaction).
As happens on such occasions, I became an instant expert on
the subject. I was appointed rapporteur to the Ethics Group of
the UK Response to the World Conservation Strategy. This led
to me serving as a member of the Ethics Working Group of the
International Union for the Conservation of Nature and taking
part in discussions with the International Environmental Law
Commission as they prepared a Covenant on Environment and
Development for the United Nations. I was one of the three UK
representatives to an Economic Nations Summit Conference on
Environmental Ethics[1] and was involved in a series of consultations
on 'a Christian Attitude to the Environment' convened by HRH the
Duke of Edinburgh at St George's House, Windsor (a follow-up to
the Assisi Declarations of the World Wide Fund for Nature, p. 209).[2]
I have been privileged to serve as President of the British Ecological
Society and of the European Ecological Federation, and have mixed
with some of the world's leading environmentalists. I gave a series
of Gifford Lectures in 1997–8 (published as *God's Book of*

[1] Bourdeau, P., Fasella, P.M. and Teller, A. (eds.) (1990). *Environmental Ethics: Man's Relationship with Nature, Interactions with Science.* Luxembourg: Commission of the European Communities. The other UK representatives were Professor Jim Lovelock, the begetter of Gaia, and Lord Nathan, a distinguished lawyer.
[2] Duke of Edinburgh and Mann, M. (1989). *Survival or Extinction.* Salisbury: Michael Russell.

Works. T&T Clark, 2003). Some of the ideas therein were put into a wider context in a book the Templeton Foundation asked me to write (*Ecology and the Environment: The Mechanism, Marring and Maintenance of Nature*. Templeton Press, 2011). This book brings together some of the background and relevance of these experiences.

Do these experiences merit a book? Emphatically not if the intention is mere autobiography. My intention is to use my particular involvement to illustrate the general situation. This book will only be worthwhile if it helps us to understand the complicated interactions of biology, sociology and morality which make us what we are and enable us better to respond to environmental challenges – which have been the subject of many committees and conferences, whose proceedings rarely make exciting reading. I cannot promise that the answers are simple. US lawyer and Yale Professor Gus Speth is on record as saying 'I used to think that the top global environmental problems were biodiversity loss, ecosystem collapse, and climate change. I thought that with 30 years of good science we could address these problems, but I was wrong. The top environmental problems are selfishness, greed and apathy.' We ignore this complexity at our peril. Our genes record past battles to survive in many environments but they (which means we) are also part of a contemporary and often unpredictable world, however much we would like to believe that we are autonomous units and can live entirely apart from our surroundings. The story recounted in the following pages is about this complexity: sometimes we can decide our own fate, but increasingly we are at the mercy of forces beyond our control and dependent on the actions of governments and corporations. This takes us ever further away from personal decision making to the often dry and seemingly dull and distant controls of national and international bodies. Notwithstanding, we are still faced with having to make decisions about our use of technology and our dependence on vested interests, however much we may fear (or perhaps welcome) being replaced by servant robots. Unpicking the reasons behind these decisions is what this book is about.

I acknowledge with gratitude those who introduced me to the ideas developed herein, in particular John Barrett of the Field Studies Council, the Revd George Bramwell Evans (Romany of the BBC), Oxford entomologist Dr Bernard Kettlewell and Canon Charles Raven of Cambridge. I thank a host of others for discussions and criticisms: Lord (Eric) Ashby, Richard Carden, Nick Clark-Lowes, Professor Ron Engel, Professor Kevin Gaston, Sir Martin Holdgate, Professor David Killingray, Dame Georgina Mace, Bishop Hugh Montefiore, Sir Jonathan Porritt, Professor Stephen Rockefeller, Dr John Sheail, Professor Chris Smout, the Reverend John Stott, Sir Crispin Tickell and many more. My thanks are also due to Dr Alan Crowden and others at Cambridge University Press: Dominic Lewis, Noah Tate and Lydia Wanstall for their tolerance and competence. I am grateful to those whose photographs are included and to Denis Lamoureux of the University of Alberta for allowing me to use his original diagram for Figure 1.3.

I have not consciously plagiarized anyone, but I must acknowledge a number of authors who I have found particularly helpful and have drawn upon:

Allen, D.E. (1976). *The Naturalist in Britain*. London: Allen Lane.

Holdgate, M.W. (1996). *From Care to Action*. London: Earthscan.

Knight, D. (2014). *Voyaging in Strange Seas*. New Haven, CT: Yale University Press.

Medawar, P. (1984). *The Limits of Science*. New York: Harper & Row.

Nicolson, M.H. (1959). *Mountain Gloom and Mountain Glory*. Ithaca, NY: Cornell University Press.

Porter, R. (2000). *Enlightenment*. London: Allen Lane.

Reynolds, F. (2016). *The Fight for Beauty*. London: Oneworld.

Sheail, J. (2002). *An Environmental History of Twentieth-Century Britain*. Basingstoke: Palgrave.

Smout, T.C. (2009). *Exploring Environmental History*. Edinburgh: Edinburgh University Press.

Wilson, E.O. (2014). *The Meaning of Human Existence*. New York: Liveright.

Wright, N.T. (2008). *Surprised by Hope*. London: SPCK.

But all this is the icing. I really ought to add Gilbert White, Alexander von Humboldt (and the wonderful biography by Andrea Wulff, *The Invention of Nature*, 2015), Charles Darwin, Charles Elton (*Ecology of Invasions*, 1958), Rachel Carson (*Silent Spring*, 1962), Keith Thomas (*Religion and the Decline of Magic*, 1971; *Man and the Natural World*, 1983), Jared Diamond (*Rise and Fall of the Third Chimpanzee*, 1991; *Guns, Germs and Steel*, 1997; *Collapse*, 2005), Simon Conway Morris (*Life's Solution*, 2005), Peter Bowler (*Monkey Trials and Gorilla Sermons*, 2007), Rebecca Stott (*Darwin's Ghosts*, 2012), Tony Juniper (*What Nature Does for Britain*, 2015) and more. But enough: of making books there is no end...

Some of the more immediate references and links are listed in footnotes. Most of these can be ignored, but I include them to confirm particular assertions or to make acknowledgement where due. Otherwise 'further reading' is listed at the end of each chapter. I have not attempted to source all my statements, but I hope I have provided enough information for anyone interested to dig as deep as they want.

1   **Choices**

You, your joys and your sorrows, your memories and your ambition, your sense of personal identity and free will, are no more than the behaviour of a vast assembly of nerve cells and their associated molecules.

Francis Crick, *The Astonishing Hypothesis*, 1995

We have a pump in our kitchen at home, connected to a deep well outside the back door. It is a relic of the time the house was built, two hundred or so years ago, when mains water did not exist. We don't use the pump. Mains water reached here in the 1880s and it is much easier and more convenient simply to turn a tap. But it serves as a reminder of a time when we had no option but to spend much time and effort coping with our environment. Most of those reading this book will take for granted piped water, easily available energy, warm houses, readily accessible transport and so on, giving only intermittent thought to the many millions without these benefits, never mind those who have voluntarily or involuntarily left their homes for squalid shanty towns or refugee camps. Notwithstanding, it seems fair to assume that a key aspiration – usually, perhaps unrecognized – of people everywhere is to cushion themselves as much as possible from the inconveniences and challenges of their surroundings, thereby effectively ignoring their environment. Is such independence from the environment a universal 'good'? We are urged to 'get out' and take exercise, told that viewing (even in pictures) the natural world may contribute to well-being – to the extent of healing from some sicknesses, both mental and physical (p. 198). What does – or should – our environment mean to us? Is it just a scenic background for our existence, a nuisance to be avoided as much as possible, or something more?

1

BOX 1.1 **The common good**

Philosophers down the ages have debated as to whether there is a 'common good' shared by and beneficial to all members of a community. Such a common good would suggest that society shares benefits by being more than a collection of individuals – that there is such a thing as 'society'. It implies that the common good of all citizens is the proper aim of government. It is a concept which lends itself to political speculation and assertions about the organization of society; it is an idea much beloved by economists, particularly in terms of their calculations of 'welfare'.

Notions of the common good are relevant to environmental actions and behaviour because of the all-too-common pursuit of what has been called 'market fundamentalism' (i.e. an unmitigated drive for growth of profits). This may lead to false accounting and endanger the 'natural capital' of the planet – the geology, soil, air, water and living things upon which we depend and from which we derive a wide range of benefits (Box 9.4 'Ecosystem services', p. 201) and which make life possible. This bias is exacerbated by the convention of measuring economic health solely in terms of gross domestic product (GDP), a metric which came into use in the 1930s and was intended as a convenient measure of economic activity. It was popularized by Maynard Keynes, among others. The problem is that GDP measures a nation's flow of income, expenditure and assets from year to year, but not its assets or balance sheet. It takes no account of any loss of natural capital nor of inequalities between or within countries. International attention is increasingly focused on the crucial significance of natural capital (see the later chapters of this work).[1] Inevitably, GDP distorts assessments if it is used as the authentic measure of financial health or social sustainability. We need to recognize from the outset that we risk both ourselves and our fellows if we concentrate on the flow of services or money and ignore the 'goods' of our environment.

[1] See also, for example, the statement 'Sustainable humanity, sustainable nature – our responsibility' produced by the Pontifical Academy in 2014 (*Science*, **345**: 1457–8). Following a policy set out in a Natural Environment

White Paper in 2011 (*The Natural Choice: Securing the Value of Nature*), the British Government receives advice on protecting and improving natural capital from a Natural Capital Committee, which reports directly to the Cabinet. Its ethos has been set out by Dieter Helm in *Natural Capital: Valuing the Planet* (2015) New Haven, CT and London: Yale University Press.

A key but neglected realization is that it is impossible to detach ourselves completely from our (multiple) environments. The most important factor is our need for energy. All energy comes from the Sun and is captured by plants; all our food comes from plants, even if it is processed through animals. Or we may get it via pickled sunlight as 'fossil fuel'. On top of that, we are born into an environment of a family after three-quarters of a year in a womb; however much we may reject social conventions, we depend on and learn from other members of our species. 'Nature versus nurture' debates rumble on interminably in various guises; the truth is that we are the product of both nature and nuture. Try as we might, we cannot simply ignore our environment.

What determines our attitudes to the environment? Do we choose them, or are determined reductionists like Francis Crick right in claiming that we are so 'hard-wired' that our choices are imaginary – nothing more than inevitable responses to stimuli conditioned by our evolutionary history (see the epigraph to this chapter)? It seems intuitively unlikely that any normal person is completely unable to make decisions for themselves; in other words, we are influenced by both environment and history. Does 'nature' or 'nurture' ultimately control our success – and decide our doom? In what sense can we make 'good' decisions – be altruistic? Are our concerns for pandas or polar bears truly meaningful, or are they merely a sort of overflow from a subliminal need for order in our surroundings?

This book is about attitudes to the environment: how they are formed and what influence they may have – or could have – for us, either as individuals or as a group. Perhaps this is a stupid quest. Only in a very few cases, where a survey has been carried out for some particular purpose, can attitudes be unequivocally ascertained. However, it seems fair to assume that an attitude manifests itself in behaviour; or, put the other way around, a person's behaviour illustrates their attitudes. Charles Darwin spent eight years devotedly studying barnacles. His commitment to barnacles was a reaction to a chance comment by his friend Joseph Hooker (p. 91). His determination to be honest with himself and his science influenced his attitude to his understanding of a particular (and very focused) part of the natural world. Behaviour showed attitude. Darwin's eight years of barnacle study may be attributed to some sort of cussedness on his part, but it seems somewhat obtuse to argue that it was predetermined by either his neurones or his DNA.

BOX 1.2   **Highland Clearances**

The eviction of farmers from tracts of the Scottish Highlands and Islands in the late eighteenth and early nineteenth century is commonly portrayed as a series of inhuman and callous acts imposed by tyrant landlords on innocent tenants (Figure 1.1). It has been condemned as racism and even genocide.

While not in any way justifying the undoubted brutality of some of those involved, historians increasingly stress that this is not the full story. Many landowners were concerned for the welfare of their people as well as for the profitability of their estates. Some bankrupted themselves in attempting to support the inexorably growing population on their territory, and in such cases their estates fell into the hands of new owners with fewer scruples or, if entailed, came to be administered by distant lawyers whose legal duty was to maximize estate income even if it meant clearing the population.[1]

The worst of the horror stories about the 'Clearances' are repeated and generalized from a handful of primary accounts. The motive of one of the most reviled landowners, the Duke of Sutherland, was not simple greed. He hoped to restructure his estate by moving

From the Painting by J. Watson Nicol   *In the Collection of Eli Lees, Esq.*

LOCHABER NO MORE.

FIGURE 1.1A Persisting attitudes. Forced evictions from parts of the Scottish Highlands and Islands by landowners converting small tenancies into large sheep farms have left an indelible prejudice to the present day. Many of those forced from their land were moved to unsuitable coastal sites or decided to emigrate.
Photo: Universal History Archive/Getty Images.

FIGURE I.IB Ruined croft houses on Vuia Mhor.
Photo © Sarah Egan (cc-by-sa/2.0).

his tenants from their unproductive hill farms to the coast, where
they could pursue fishing or weaving, or work at the coal mine, salt
pans or brick works in which he had invested. His plans, however,
were ill thought-out, with no immediate provision of suitable
accommodation for the displaced people, and took no account of the
difficulty of impoverished subsistence farmers becoming successful
fishermen. The worst cruelties on his estate were not perpetrated
by the Duke himself but by his overambitious agent, Patrick Sellar,
who stood to gain personally from the evictions by taking over
some of the vacated land for himself.[2] The Duke's plans failed badly;
they have been described as a 'typical example of social engineering
which met neither the hopes of the benefactors nor the needs of the
beneficiaries, but produced social disaster'.

Much of the Scottish Highlands is wet blanket bog. It used to be
thought that the area was a 'man-made wet desert' produced by forest
clearances in medieval times, followed by subsequent overgrazing.
This is wrong. Blanket bog is the normal climax vegetation for
the area. Woody remains in the bogs are of trees dated to around

4000 years ago when the bogs were forming at a time of climate change in the Bronze Age. There is no clear evidence of damage by humans in the Middle Ages, although heavy grazing since the eighteenth century by cattle and then by sheep has prevented natural regeneration in the drier areas.[3]

The human population in much of the Highland area more than doubled in a generation from the middle of the eighteenth century, largely because the introduction of potatoes increased the yield of poor land and allowed better survival of the inhabitants, albeit with a low standard of living. At first, the lairds welcomed this population increase because it helped their recruitment for the Highland regiments and their own rank as officers in the army itself, but this ended in 1846 with the failure of the potato crop and the consequent famine. How were the lairds to 'improve' the lot of their people? The population density was well above the carrying capacity of the land. The rate of bankruptcies among the lairds themselves was increasing. Some of the more compassionate landowners sought to provide for their people by helping them to emigrate. In other cases, the subsistence farmers were displaced from their traditional homes, and resettled in marginal land around the coast. A few ended up in model fishing towns like Wick, Ullapool or Bowmore, but the episode has left an abiding bitterness and angst. There were certainly instances of extremely cruel behaviour by those with power, but the impact of the Clearances has been much overstated. Many landowners did their best for their people, some of them paying the costs of emigration to North America. The population in the Highlands fell by a similar amount to that in many parts of the rural Lowlands and only by a third of that in Ireland during the same period (without including the deaths of a million people who starved in the famine in Ireland). By 1880, Clearances were illegal, but they had largely ceased by then.

The legacy of the Clearances is a strong folk memory of maltreatment, a memory which still dominates and rankles, and determines attitudes to authority and land ownership in the Highland area. They precipitated an evolution in agricultural practices which had been taking place for more than a century in England, with greater landlord control, professionalization of management, shorter leases, extension of enclosures and the erosion of small farms. The unrest from these changes was magnified because they came all at once in the Highlands. Environmental attitudes in this episode have been more shaped and fixed by emotion than by historical fact. There is a proper and understandable passion for place, but belief in an illusory golden age has been fostered by memories of hardship and starvation.

[1] Richards, E. (2016). *The Highland Estate Factor in the Age of Clearances*. Lewis: Islands Book Trust.
[2] Grimble, I. (1962). *The Trial of Patrick Sellar*. London: Routledge and Kegan Paul.
[3] Smout, T.C. (2000). *Nature Contested: Environmental History in Scotland and Northern England Since 1600*. Edinburgh: Edinburgh University Press.

BOX 1.3 **Building the railways**

When the railway network was spreading in the mid-nineteenth century, there was much local opposition ('Not in my backyard'). William Wordsworth did his utmost to prevent the building of the line between Kendal and Windermere. He contrasted nature with the world of materialism; he wrote 'Because we are insensitive to the richness of Nature, we may be forfeiting our souls.' He condemned the proposed line as 'Utilitarianism, serving as a mask for cupidity and gambling speculations.'[1]

In 1863 the Midland Railway was widely criticized for despoiling the natural beauty of the Peak District with the line it was building between Derby and Manchester, in particular a five-span 91 m long viaduct at Monsal Head over the River Wye. John Ruskin wrote:

> There was a rocky valley between Buxton and Bakewell once upon a time divine as the Vale of Tempe... You Enterprised a Railroad through the valley – you blasted its rocks away, heaped thousands of tons of shale into its lovely stream. The valley is gone, and the Gods with it; and now every fool in Buxton can be in Bakewell in half an hour, and every fool in Bakewell in Buxton.[2]

The line closed in 1968 as part of the Beeching era cuts (Figure 1.2).

FIGURE 1.2 A change in attitude. The 91 m (300 feet) long Monsal Head Viaduct built in 1863 to carry the railway from Matlock to Buxton and Manchester. It was reviled by contemporary environmentalists. After the rail line was shut in 1968, the intention was to demolish the viaduct – but it was 'saved' by the vehement protests of walkers and scenery lovers who argued that it significantly enhanced the view; it is now a Grade II listed monument. Photo © Eleanor Scriven.

There were proposals to demolish the viaduct, but as a result of a public outcry, it was acquired by the Peak National Park Authority and is now a listed structure. It is now generally regarded as a thing of beauty; attitudes towards it have changed radically.

[1] Wordsworth, W. (1844). Kendal and Windermere Railway: two letters reprinted from The Morning Post. In: *The Prose Works of William Wordsworth, Volume 2*. London: Edward Moxon, Son and Co.

[2] Ruskin, J. (1871). *Fors Clavigera: Letters to the Workmen and Labourers of Great Britain*. Orpington: George Allen.

Reading attitudes from behaviour may be questioned, but one thing is certain about attitudes: they have changed radically over time. Put crudely, we have moved from a wholly animal existence to a much more sophisticated one. Not many people nowadays have to live in constant fear of being eaten by a lion or a bear. Food shortages still exist, but there are agencies to note and alleviate them. The factors which determine or change our attitude are complex. Sometimes they are the result of a catastrophe (drought, tsunami, pollution, disease, earthquake), or legislation conceived for whatever reason (disease control, safety, animal welfare, planning constraints, vested interest), or technological innovation (efficient sanitation, easily available transport, user-friendly identification guides, satellite imagery). At other times a lever of change may be the influence of a particular person (John Locke, Samuel Taylor Coleridge, John Muir, Teddy Roosevelt, Charles Darwin, Rachel Carson, Julian Huxley, David Attenborough – the list is long). This book explores these levers and how they are perceived. It is necessarily 'broad-brush' and therefore ludicrously incomplete. Only a complete history of humankind would reveal all the controls, and this book does not pretend to be a history. Indeed, it may infuriate by deviating from a formal time-line. By being selective, I am giving a hostage to critics who will always be able to find influences I have not included and are important in their judgement.

One other caveat: my starting point is the nature of things (their ontology) – science and the effects of scientific thinking. I make no apology for this: I am a natural scientist. My expertise, such as it is, is in studying the structure and functioning of the natural world, particularly its living animals and plants. My defence is that our place in this world is a logical starting point: our being and surroundings are ultimately determined by the interactions of physical realities – animal, vegetable and mineral – stretching from molecules and chromosomes to meteorology and cosmology. These interactions are the building blocks for us as individuals but also for society and its behaviours, and our understanding of them. Although we will have to stray into philosophy, emotion and religion, we need to accept that these disciplines are dependent on and secondary to the underlying mechanisms.

Harvard biologist Ed Wilson adds authority to this context. He writes:

> The task of understanding humanity is too important and too daunting to leave exclusively to the humanities. Their many branches, from philosophy to law to history and the creative arts, have described the particularities of human nature back and forth in endless permutations, albeit laced with genius and in exquisite detail. But they have not explained why we possess our special nature and not some other, out of a vast number of conceivable possibilities. In that sense the humanities have not achieved nor will they ever achieve a full understanding of the meaning of our species' existence. So, as best we can answer, just what are we? The key to the great riddle lies in the circumstance and process that created our species.[1]

The more sensitive may recoil in horror from the statement attributed to Bertrand Russell that 'science is organized common sense; philosophy is organized piffle' (perhaps taking heart from Plato's criticism of Athenian politicians, that 'rudeness is taken as a mark of sophistication'), but science is where we need to begin, testing any conclusions against ontological reality rather than relying upon secondary inference.

BOX 1.4 **The beginning of science**

Science in the modern sense began in 1543. Two books were published in that year which laid its basis – at least for science as understood in the West. Both affected the way information is processed and choices made. The two books were *De Fabrica Humani Corporis* by Andreas Vesalius, based on careful dissection of the human body, which refuted hallowed assumptions based on tradition rather than observation (such as the mythical Bone of Luz

---

[1] Wilson, E.O. (2014). *The Meaning of Human Existence.* New York: Liveright, p. 17.

from which the resurrection body was supposed to grow, or Adam's 'missing rib' from which Eve was made); and *De Revolutionibus Omnium* by Nicolaus Copernicus, which argued that it made more sense of astronomical observations if one assumed that the Earth moved round the Sun rather than remaining fixed at the centre of a rather small cosmos. These two books laid the basis of experimental science, based on deduction from the existence of testable realities. Copernicus showed that traditional ways of understanding the cosmos were not the only or necessarily the best ways; Vesalius destroyed the myths enshrined in biological and medical texts by the authority of Galen and other ancient authorities, and laid the ground for modern scientific rigour (Figure 1.3).

Does all this matter? We can acknowledge that we all benefit from science and depend on its products in food, transport, clothing, medicine – in virtually every aspect of human existence – but it is hard to get excited about the intellectual achievements of half a millennium ago. But there is a legacy of 1543 which only slowly unravelled and even today is commonly ignored in practice. For most of our history, we – humankind – assumed that we were at the centre of things: that we were uniquely privileged and able to draw upon the apparently unlimited abundance of the world in which we lived. The world was made for us. We were created as its rulers, responsible perhaps to a benevolent although judgemental God, but otherwise restricted only by the social constraints of our particular surroundings. 1543 changed all that – or at least, began to change it. We have had to accept that we are the denizens of a smallish planet in an enormously great space. No longer could we relate to a God who lived above a rather solid sky (made from something like copper sheeting, with holes to let the rain through). Increasingly we have been forced to recognize that the world was much, much older than a few thousand years. And then Darwin clinched our angst by producing evidence that we are not as distinct as we had liked to believe, that we are in fact animals.

Some years ago a Nobel laureate, Peter Medawar, wrote *The Limits of Science*,[1] arguing that there are no discernible limits to the power of science to answer those questions which it is capable of

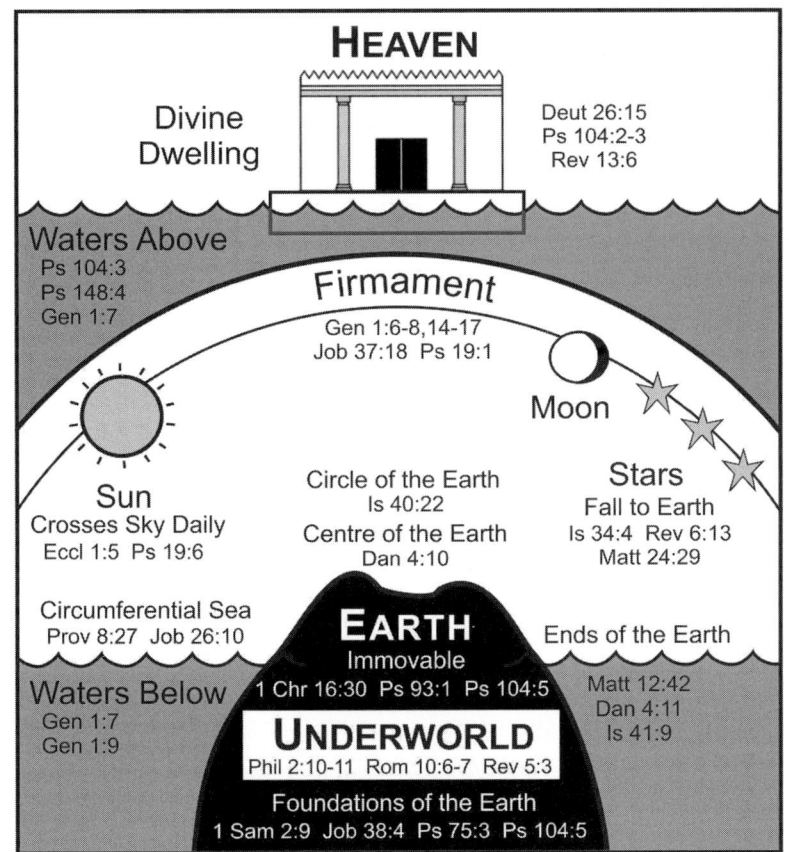

FIGURE 1.3 Mistaken attitude. Three-tier universe as understood by the ancient Israelites on the basis of the Hebrew Bible (i.e. Old Testament). Drawn by Denis Lamoureux.

answering – those where we can suggest and investigate the cause(s) of some phenomenon. But he unapologetically asserted that there are some questions that science cannot answer and will never be able to answer. He said that for these we must seek what he called 'transcendent answers', which he defined as 'answers that do not grow out of our need to be validated by empirical experience, answers that belong to the domains of myth, metaphysics, imaginative literature or religion'.

This was not an attempt to smuggle religion in by the back door and certainly not to spite nouveau atheists by thwarting their attacks before they had made them. According to Medawar's wife, the purpose of her husband's book was 'that science should not be expected to provide solutions to problems such as the purpose of life or the existence of God for which it is unfitted'. Indeed, Medawar himself confessed 'I regret my disbelief in God and religious answers generally, for I believe it would give satisfaction and comfort to many in need of it if it were possible to discover and propound good scientific and philosophic reasons to believe in God'.[2] This makes much more significant his recognition of the limits of science as normally practised and his acceptance of the reality of 'metaphysical questions'.

[1] New York: Harper & Row, 1984.

[2] Medawar, P. (1996). *The Strange Case of the Spotted Mice*. Oxford: Oxford University Press..

There are those who infer that our discoveries in science show we are bound and constrained in the same ways as any animal, that we are no more than 'naked apes' – or perhaps wonderfully hedonistic robots. This seemed to have been Francis Crick's view. Is this fair? Are we apes on the way up? Are we merely DNA factories driven by deterministic physico-chemical reactions, helpless pawns in a post-modern jungle? Or alternatively, is there any credibility in a belief that we are somehow embodied souls, as our forebears assumed?

Neuroscience has disabused us of any idea that we are a tripartite compound of body, mind and spirit, or even a binary amalgam of body and soul, but there is no easy way out of J.B.S. Haldane's puzzle when he wrote in a 1927 essay 'When I am dead':

If my mental processes are determined wholly by the motions of atoms in my brain, I have no reason to suppose that my beliefs are true. They may be sound chemically, but that does not make them sound logically. And hence I have no reason for supposing my brain to be composed of atoms. In order to escape from this necessity of sawing away the branch on which I am sitting, so to speak, I am compelled to believe that mind is not wholly conditioned by matter. But as regards my own very finite and imperfect mind, I can see by studying the effects on it of drugs, disease, and so on, that its limitations are largely at least due to my body. Without that body it may perish altogether, but it seems to me quite as probable that it will lose its limitations and be merged into an infinite mind or something analogous to a mind which I have reason to suspect probably exists behind nature. How this might be accomplished, I have no idea.[2]

BOX 1.5   **Hereditary genius**

In the almost aggressive contemporary fashion for equality, it may seem disingenuous to raise the question and significance of inequality. In 1869 Charles Darwin's cousin, Francis Galton (they were both grandsons of Erasmus Darwin), published *Hereditary Genius: An Inquiry into its Laws and Consequences*, listing many families where particular talents spanned a number of generations. For example, there were eight generations of musical genius in the Bach family over 250 years. The nineteenth century produced several very distinguished scientific families. There were the Darwins, and also the Huxleys and the Haldanes.

The Haldanes were a long-established Scottish family, with an estate at Gleneagles. A mid-eighteenth-century Haldane made a fortune as the commander of an East India Company ship. Another Haldane of the same era was the ambassador to Russia. However, the most prominent Haldanes at the end of the century were two brothers (Robert and James, born in 1764 and 1768, respectively), who after naval service became noted evangelists and benefactors of the Scottish Church in the Highlands, on occasion in company with the Cambridge-based Charles Simeon, often credited with rescuing the English Church from the cynical rationalism of

---

[2] In *Possible Worlds*. London: Chatto & Windus.

the Enlightenment.[1] James's half-brother served as President of the Royal College of Physicians of Edinburgh, while two of his grandchildren were noted philosophers: the older one (Richard) was well known as a reforming politician and then Lord Chancellor, becoming Viscount Haldane of Cloan; the younger (John) was a respiratory physiologist, who worked with his mother's brother in the Physiology Department at Oxford.[2] (Their uncle, John Scott Burdon-Sanderson, was the first Professor of Physiology in Oxford. He was the person who named ecology as a distinct discipline – p. 146). A sister was a senior nurse, appointed as a Companion of Honour. J.B.S. (Jack) Haldane (1892–1964) was a son of John the physiologist. As a schoolboy, he worked with his father on breathing problems of deep-sea divers, submarine workers and coal miners. He fought in the First World War and during it (in 1915) published with his sister, the novelist Naomi Mitchison, the first evidence of genetical linkage in mammals. He subsequently had a distinguished academic career in biochemistry and genetics, besides an active political career as a Marxist and science populist.[3] Both Jack and his father were Fellows of the Royal Society, as were two of Naomi's sons and a grandson.

It would be wrong to make any generalization whatsoever from the Haldane family – or any other family. The concentration of talent and achievement in the Haldanes does not prove anything about either hereditary genius or nurture. Notwithstanding, the span from unthinking imperialist to fervent evangelist and then to hard-headed critical scientist is intriguing. It certainly does not imply the sort of autonomy suggested by Francis Crick. We are all different. We all have talents. The challenge is not to be a Haldane or Crick, but to find our personal talent and fit it to its appropriate hole.

[1] Haldane, A. (1853). *Memoirs of the Lives of Robert Haldane of Airthrey and of His Brother James Alexander Haldane*. London: Hamilton, Adams, and Co.

[2] Goodman, M. (2007). *Suffer and Survive*. London: Simon & Schuster.

[3] Clark, R. (1968). *J.B.S.: the Life and Work of J.B.S. Haldane*. London: Hodder & Stoughton.

Tom Stoppard wrestled with the same dilemma in his play *The Hard Problem*, asking whether consciousness is really nothing more than a by-product of neurological activity. He explored the assumption that body and mind are separate 'things', and found it necessary to reject a rigid distinction between the two. He used game theory to suggest that David Hume was right in thinking that cooperation is the result of people helping those who aid them and punishing those who punish them. He quotes the enigma of arch-reductionist Richard Dawkins confessing to be 'a sort of dualist by default since goodness is not the same as good behaviour'. Stoppard admits a similar uncertainty in the programme for the play: 'If dualism were true, what would we notice as being different around the world we inhabit? We might survive bodily death, for example, although it would be hard to notice that and tell the tale. What else? I suppose as an evolutionist, what really bothers me about dualism is the question of when, in evolution, the mind started creeping in.'

Stoppard's confession is honest and certainly more realistic than yet another, albeit very dated perception of science and scientists as portrayed by Ludwig Büchner in his book *Kraft und Stuff* (published in 1855; translated into English as *Force and Matter*). It went through twenty-one German editions and was translated into fifteen languages. The book itself is now largely forgotten, but it has had an enormous and persisting influence. Büchner argued that matter was all; life and humans have emerged as unplanned and haphazard accidents of nature. He was the intellectual grandfather of Francis Crick, Richard Dawkins, Daniel Dennett and Christopher Hitchens. Notwithstanding, Owen Chadwick condemns Büchner's belief as being nothing more than glib assumptions about science. He comments, 'The great and lasting consequence of Büchner's work was the expression of popular superficial thought influenced by the successes of the natural sciences.' The problem is eradicating wrong ideas when they take hold. Chadwick asks, 'How influential is a popular book? Bishop Robinson wrote *Honest to God* and is said to have sold millions of copies. Does it follow that the book changed so many people's minds? Is it possible that not everyone who bought the book understood it, and perhaps not even the author understood every word of what he had written?'[3]

Do beliefs and behaviours really change as a result of such books as Büchner's or Robinson's? Or Stephen Hawking's *Brief History of Time*? Gilbert White's *Natural History of Selborne* and Rachel Carson's *Silent Spring* have certainly had significant and lasting effects. *Kraft und Stuff* appeared four years before *On the Origin of Species* and five years before the storm of shocked doubt precipitated by the theological collection of *Essays and Reviews*. All these produced intellectual trauma in the chattering classes. We cannot ignore such landmark writings – nor what followed from key people or subsequent actions (or legislation). Clearly, inferring attitudes is an imprecise activity; hopefully identifying some historical event or writing may help, while recognizing that such inferences may be overinfluenced because of a serendipitous document or a particular charismatic advocate.

Along with Haldane and Stoppard, I can only speculate about what may happen after death. None of us has any privileged information about this, despite proselytising enthusiasts who extrapolate from a 'near-death experience'. But thinking about life (and death) brings us into the world of values and attitudes, moving us from natural to social science. It opens a Pandora's box of choice-making – ranging from the possibility of opting out of any decisions and being buffeted by one's surroundings to actively seeking positive ways forward; between submission and reaction.[4] The possibility of making choices means we can make wrong ones as well as good ones, and introduces the hazards and privileges of responsibility for our decisions – responsibilities which grow as our decisions affect people

---

[3] Chadwick, O. (1975). *The Secularization of the European Mind in the 19th Century*. Cambridge: Cambridge University Press, p. 172. *Honest to God* was a book published by SCM Press in 1963.
[4] Wright, T. (2008). *Surprised by Hope*. London: SPCK.

other than ourselves, never mind our children and grandchildren. Our attitudes to the environment may have considerable significance.

Social scientists are wont to scoff about natural scientists trespassing, as they see it, on their territory. They tend to be fiercely and often arrogantly possessive of their domain, but by definition they deal with second-order interactions – social factors are necessarily based on the primary meeting of individuals with a frequently dangerous world. But there are real decisions to be made. Do we put saving pandas above producing food for an ever-increasing human population? How do we measure the importance of wilderness against wind turbines or oil pipelines? Am I prepared to become a vegetarian and release more land for cereal crops rather than feeding animals? Is it right to sacrifice environmental care to poverty alleviation? Do we have to choose between the two? How do we value our environment? We certainly have to make decisions and these decisions are likely to have more than pure academic interest.

What are the bases for all these decisions? The obvious place to seek them, as the King of Hearts said to Alice, is to 'begin at the beginning and go on to the end and then stop'. This book is an attempt to begin at the beginning. How near we get to the end is for the reader to judge. There is no intention (or pretension) to present an original thesis; the aim and hope of this book is much more modest – to bring together ideas and information from different fields and contexts so as to give its readers a firmer basis for making their own choices.

FURTHER READING

Berry, R.J. (ed.) (1993). *Environmental Dilemmas*. London: Chapman & Hall.
Green, J.B. (ed.) (2004). *What About the Soul?* Nashville, TN: Abingdon.
Haldane, J.B.S. (1927). *Possible Worlds*. London: Chatto & Windus.
Knight, D. (2009). *The Making of Modern Science*. Cambridge: Polity.
Medawar, P. (1984). *The Limits of Science*. New York: Harper & Row.
Wilson, E.O. (2014). *The Meaning of Human Existence*. New York: Liveright.

2   No Primeval Eden

We tend to have illusions about our past: the sun was always shining (or the snow was always perfect for skiing); the champagne was properly chilled, the beer the right temperature; the daughter's boyfriend a budding Einstein crossed with Eric Liddell or David Beckham. But we know this is not true. Time is neither inevitably progressive nor necessarily regressive, but it is unforgivably remorseless. There has been no Golden Age either in our personal past or further back for any of our forebears. There never was a Paradise when wolves lay down with lambs and where poisonous snakes, biting midges, clothes moths, drug dealers and loan sharks did not exist. There was no primeval Eden, occupied by a single couple a week or so after the world came into being. Even more daunting, life itself may not be inevitable. For a billion years there was no life of any sort on this planet Earth.

Our planet was formed around four and a half billion years ago. Primitive self-replicating organisms emerged around 3.9 billion years ago. Further development was slow. It was half a billion years ago during the time of the 'Cambrian explosion' that a change from simple, sedate animals – non-motile bacteria-like cells – to highly mobile animals took place. It is likely that this was the consequence of an increase in atmospheric oxygen – an environmental change. Primitive fish began to swim around the oceans; not long afterwards, in terms of geological time, some of them began to breathe air and survive on land. Things then speeded up. Dinosaurs appeared rather less than a quarter of a billion years ago and lasted 170 or so million years, disappearing sixty-six million years before now (apparently by *force majeure* – perhaps as a consequence of a huge meteorite hitting the Earth, perhaps due to a burst of volcanic activity, perhaps both),

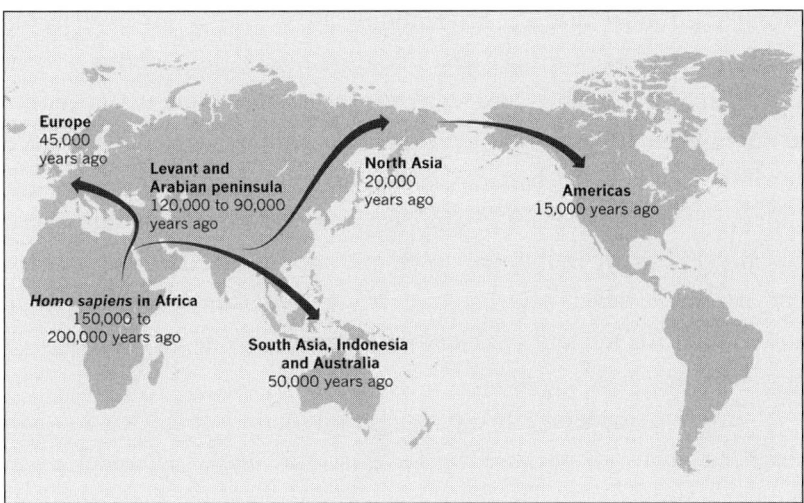

FIGURE 2.1 The timings of colonizations following the spread of humans from their origin in Africa, using fossil, archaeological and genetic evidence.
Reprinted by permission from Macmillan Publishers Ltd from Peter deMenocal and Chris Stringer (2016). Human migration: climate and the peopling of the world. *Nature*, **538**: 49–50.

leaving space for mammals. Fossils of the earliest primate so far found are dated fairly soon after that, but the first human species (*Homo habilis*) arrived a mere (in geological terms) two and a half million years ago. Our own species (so-called archaic *Homo sapiens*) emerged in Africa between 400,000 and 250,000 years ago, with anatomically modern humans appearing 200,000 years ago. There were several emigration events out of Africa (perhaps driven by climatic changes), but the most significant move for the lasting colonization of Eurasia began around 60,000 years ago (Figure 2.1).

The history of all this developing life involved conflict – not a series of battles fought with arrows or attack helicopters; rather a procession of long-drawn and attritional sieges, punctuated at intervals by attempts to escape from the inimical pressures exerted by an all-pervasive environment. Such battles haven't ended: amphibians throughout the world are currently dying at an unsustainable rate through a fungal disease, chytridiomycosis; white nose syndrome is a major scourge for bats; the myxomatosis virus remains a virulent affliction in rabbits; red squirrels in Britain frequently succumb to a pox virus carried by the closely related greys without apparent detriment; cultivated bananas are seriously at risk from the fungal Panama Disease; diseases in elm, ash, oak and horse chestnuts periodically kill both individual trees and whole forests. Humans are not and have never been exempt from such campaigns. We have had to fight for life and food just as determinedly as an oak tree or a mosquito or *Tyrannosaurus*.

We live in a world at war, fought on many fronts. Our survival has been achieved because we have (so far) been successful in it. We have attained progressively greater independence from environmental assaults, but no single battle has been decisive; new challenges continually arise. And dangerously, success can too easily lead to complacency. We know in principle how to deal with many of the afflictions that faced our ancestors – cold, hunger, wild beasts – but we are unlikely ever to triumph completely. New diseases are continually arising – HIV, SARS, MERS ('camel disease'), Ebola, hepatitis C, Legionnaire's disease, Lyme disease, 'mad cow' disease – and the bacteria we thought were beaten are fighting back, as strains become resistant to more and more antibiotics. And on top of attacks from within (as it were) are challenges from outside: earthquakes, climate uncertainties, prophecies of food, water and energy shortages, never mind political instabilities and global terrorism. A previous Chief Scientific Adviser to the British Government, Sir John Beddington, has warned of an approaching 'perfect storm' around 2030, since we will have to produce 50 per cent more food and energy and 30 per cent more fresh water for a growing population, while at the same time mitigating and adapting to climate change (Figure 2.2). It seems unlikely that we will be able to avoid the eye of this particular hurricane.

Perhaps the difference as far as we are concerned is that the weapons have shifted more for us than for other creatures.

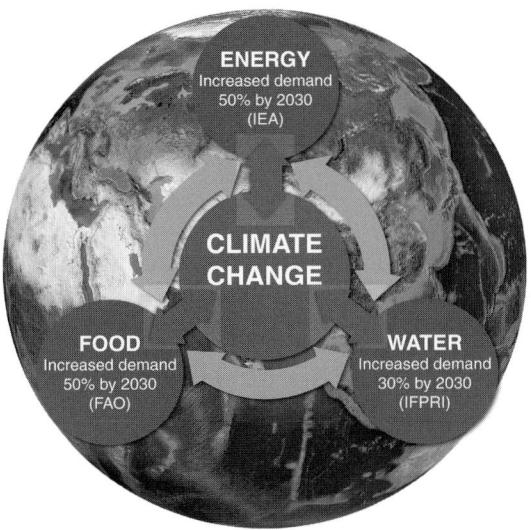

FIGURE 2.2 We are facing a 'perfect storm' with an inexorably growing
population which will need 50 per cent more food and energy and 30
per cent more fresh water over the next generation, at the same time
having to adapt to climate change.
A scenario drawn up by former government Chief Scientist,
Professor Sir John Beddington. Reproduced with permission from
Beddington, *Biodiversity: Policy Challenges in a Changing World*.
London: Government Office for Science 2009.

Notwithstanding, Darwin was brutally realistic when he reminded
us at the end of his massive *Descent of Man* that despite 'all our
noble qualities, with sympathy which feels for the most debased,
with benevolence which extends not only to other men but to the
humblest living creature, with [a] god-like intellect which has pen-
etrated into the movements and constitution of the solar system –
with all these exalted powers – Man still bears in his bodily frame the
indelible stamp of his lowly origin'.[1] Like it or not, we cannot escape
our 'lowly origin'. We share a history with all living creatures.

An inescapable problem is that the pressures on us repeatedly
change, often without us noticing. No sooner does one hazard lessen

[1] London: John Murray, 1871.

than another presses. Twenty-first-century stresses are very different from those we faced when we left the other apes and first became human. But nothing is completely new. We currently face climate change, with wildly different perceptions of what it may mean for us. The first humans also faced climate change, although in retrospect it turned out to be a massive boon for them (and us). What happened around six million years ago was that the tropical latitudes of East Africa where our predecessors lived were becoming drier, and the friendly neighbourhood forest was shrinking. Our closest cousins, the chimpanzees, retreated with the forest. The human (or really, pre-human) route to survival seems to have been the opposite; we migrated to the coast and shallow sea, where there was easily available and nutritious food but where we had to stand on two legs. Of course those who took this route did not make a conscious decision; their choices would have been determined solely and starkly by the possibility of better survival (and therefore reproduction). If that ancient climate had not changed, we might still be living in trees in tropical Africa.

The idea of an aquatic phase in our history was propounded by Oxford professor Alister Hardy. He suggested that we lost our body hair like whales, but kept hair on our head to protect us from the Sun. Standing upright meant the pharynx descended from the nasal area to the throat, perhaps facilitating the control of speech – a key factor in humanness (see p. 33).[2]

Hardy's ingenuity is contentious (it is said that he dreamt up the idea as the topic for a lecture to the British Sub-Aqua Club); our early history can be little more than guesswork. Nevertheless, some things are certain: there is no real doubt that the human line originated in Africa, spread from there on one or probably more occasions between 1.8 million and 1.3 million years ago, and then again more recently, as noted above. Cambridge professor Tim Clutton-Brock, who knows more about animal behaviour than most people, has

---

[2] Morgan, E. (1972). *The Descent of Woman*. London: Souvenir Press.

argued that the human path was made possible by a newly arisen ability to get on with each other, helped by being able to use precise and flexible language. He points out that:

- apes do not cooperate well with each other, so it is unlikely that the common ape/human ancestor would have been a cooperator;
- human cooperation is unusual in being highly developed between non-related individuals;
- pairs of non-human animals cannot make agreements because they have no language to achieve this; such agreements only became possible once language had developed;
- agreements between pairs would have rapidly developed into simple social contracts enforced by group ideology (including religion);
- if all this is correct, it implies that the evolution of human cooperation (and humanness itself) may be a relatively recent phenomenon (i.e. within the last million years).

Southeast Asia and southern China seem to have been the first regions colonized by 'modern' humans migrating from Africa. Europe (and, indeed, northern China) was occupied more recently, around 40,000 years ago, and the Americas perhaps only fifteen thousand years ago. Colonists spread across the Bering Strait into Alaska (Figure 2.1) and then took only a thousand years or so to reach Patagonia. Ethnic differentiation between modern human populations (so-called but mislabelled as 'races') is therefore evolutionarily recent, a result of divergent evolution between geographically separated groups during the last 50,000 to 100,000 years.

Why modern humans reached Europe later than they reached China and Australasia, which are geographically much further from their starting point in Africa, is not clear. Perhaps as a tropical species they spread eastwards in warmer climes rather than into colder northern regions. Possibly our first cousins, the Neanderthals, who occupied much of Europe between 200,000 and 35,000 years ago, hampered our northward spread. Certainly we coexisted and inter-bred to some extent with the Neanderthals in Europe from around 45,000 years ago, until our Neanderthal relatives went extinct more than 5000 years later. The Neanderthal sub-species (*neanderthalensis*) was sturdier than the one that survived (our ancestors, *sapiens*). Darwin argued in the *Descent of Man* that 'the art of making fire is probably the greatest [discovery], excepting language, ever made by man'. The ability to use and control fire for keeping warm and for cooking dates back at least 800,000 years and is obviously highly important, but both Neanderthals and *sapiens* learnt to use fire, although excavated Neanderthal hearths show that their meat was often eaten raw.

There is no evidence that we actively killed off our cousins, nor that we somehow overcame them by some sort of stealth. The expert consensus is that the reason the Neanderthals lost their last battle was because they just could not cope as well as us with the arrival of the last Ice Age. Analysis of Neanderthal diet by examination of their bones show that they seemed to depend mainly on meat from large game, such as reindeer, mammoth, bison and horse. In contrast, our *sapiens* ancestors ate more fish and birds, particularly water fowl. As a result the Neanderthals may have been more affected than *sapiens* by the changes in fauna as the climate deteriorated. In addition, they did not develop their stone hunting tools as efficiently as *sapiens*. It seems not unreasonable to assume that they failed because of a simple inability to adapt. We know too little about the other races (sub-species or even species) of *Homo sapiens* (*floresiensis* and the Denisovans, perhaps others) which coexisted with *sapiens* to speculate about the reasons for their lack of success.

What is beyond dispute is that our survival was – and is – inextricably linked to how we coped and reacted to our environment. Nowadays our food is more likely to come from Walmart or Sainsbury's than migrating buffalo or wild fruit, but for most of our history we have had to be energetically proactive in finding our food. The first animals we domesticated were dogs, bred from wolves probably more than 15,000 years ago, perhaps to provide a source of easily available food. This seems to have occurred independently in western Europe and in China. It may well have been that these early domesticates were found to be useful also in herding and thence taming wild cattle, providing the basis of the origin of settled farming we call the Neolithic revolution. However, the most important centre of domestication of both animals and plants was in the 'fertile crescent' around the eastern Mediterranean, a place which supported a rich mixture of African and Asian species. The region was home to the progenitors of some of our most important cultivated crops: emmer wheat (*Triticum dicooccum*), einkorn (*T. monococcum*), barley (*Hordeum vulgare*), flax (*Linum usitatissimum*), chickpea (*Cicer arietinum*) and lentil (*Lens culinaris*). The dogs were presumably used for hunting and herding in an area where four of the five animals which were the next candidates for domestication lived – goats, sheep, cows and pigs (the fifth species – the horse – lived not too far to the east). The positive advantages of staying in the same area, growing crops and herding animals were such that it led to the beginning of farming, which meant settled communities, and in them the possibility of specialized occupations within the community.

What about the humans themselves? In its early stages the human line must have undergone a number of significant genetical changes, although it is worth noting that our gene complement differs from that of the chimpanzees by only about 4 per cent of the total genome – as little, for example, as between an Indian and an African elephant, or a horse and a zebra. Indeed, the likeness is so great that it has been suggested a pedantic taxonomist might classify us as a third chimpanzee species, alongside the common and pygmy chimps. One difference is that we have twenty-three pairs of chromosomes, while all the other apes have twenty-four pairs. However, it is easy to see under the microscope that this difference is due to two ape chromosomes joining together to form one long chromosome. This would not have led to a major loss of genes. Most of the differences between the human and chimp genomes are the result of small chromosomal changes (deletions and insertions in individual chromosomes). This is not to imply that we are nothing more than a sort of chimpanzee, but it shows the need to recognize the subtlety of the distinctiveness between us.

An important differentiator is probably the gene *PRDM9* (which uninterestingly stands for 'protein domain zinc finger protein 9'), which affects whether hybrids can easily form – or not; it has a very different sequence in humans and chimps. Genes which affect brain growth also show greater differences between the human and chimp genomes than ones affecting other organs. Clearly, our humanness has a genetic basis, but many significant differences between the chimps and us seem to be epigenetic ones – that is through the regulation of gene expression (affecting how mRNA and other DNA products are produced) rather than the genetic code itself.

Signs of our distinctiveness as humans first appeared in Africa about three million years ago with evidence of meat-eating and tool-making. Around the same time our brain increased in size relative to the rest of the body. All areas of the brain grew, but the cerebral hemispheres (the thinking areas) increased proportionately more than the cerebellum (which regulates bodily control mechanisms). The speed of cerebral enlargement was greatest at a time of apparent increase in behavioural complexity in both Neanderthals and early modern humans. During the same period, our rate of physical development also changed. Humans take twice as long to mature after birth as chimpanzees or gorillas, although our gestation times are similar. For example, the human primary dentition (our 'milk teeth') emerges at around six months and the secondary ('permanent') dentition at about six years, contrasting with three months and three years respectively in chimpanzees. Humans grow for twenty years and live for seventy years, whereas the timings for other great apes are ten and thirty-five years. Sexual maturity in human females takes place at twelve years of age; it is six to eight years in the apes. Intriguingly, this change may be associated with a mutation that also affects fat metabolism – which could be a reason why we get fat but apes do not. Mutations producing 'neotenous' (rate of development) changes are known in many organisms. The best known is probably the salamander, the axolotl, which becomes sexually mature while retaining many of the traits of a newt larva. Studies of fossil skull size and tooth eruption order indicate that the change in growth pattern took place in *Homo erectus* about a million years ago.

The large brain of humans is possible because it is able to grow after birth: the human skull is 'rubbery' at birth and only ossifies during the first year. This means the skull can squeeze through the mother's birth canal but permits the brain to continue to increase in size – which it does for a year or so, as opposed to only a month or two in chimpanzees. However, large brains require large amounts of energy: about 20 per cent of resting metabolism in humans is needed to support brain metabolism, compared with 9 per cent in chimpanzees and 2 per cent in marsupials (kangaroos and their relatives). Larger brains have a low tolerance for variations in temperature, blood pressure and oxygenation. This means there must have been acute pressures (and therefore strong natural selection) to increase energy intake to tighten our physiological regulation. The likelihood is that a shift from vegetarianism to meat-eating was crucial in the survival of this large-brained line; hominids feeding on animals could significantly increase their caloric intake much more easily than fruit-eaters like chimpanzees. The occurrence of animal bones and stone tools at early *Homo* sites is probably an indicator of this, as is the flatter face and reduction in size of our molar tooth at the same time, consistent with a reduction in feeding on tough and fibrous foods. Hunting is more efficiently carried out cooperatively, so this in turn would be likely to encourage social grouping for defence and food storage and an increase in male–female bonding for increased stability.

None of this shows what humanness really is. Should we define it by some developmental feature or is it better marked by behaviour?

At one time it was assumed that standing on two legs made us human. Then it was suggested that using a tool meant humanness – until it was realized that birds and even some insects may use tools, never mind other primates. For the ancient Greeks, humanness was the ability to reason, linked to having an immortal 'soul'. For others, humanness is shown by formalized burials, implying a belief in life after death; by beads or body ornaments; by cave paintings; by findings of musical instruments; or by the origins of agriculture and animal domestication. Some of these components of 'humanness' can be recognized in the African archaeological record in the Middle Stone Age at least 40,000 to 50,000 years ago (stone tools, specialized hunting, use of aquatic resources, long-distance trade, art and decoration). The evidence for such markers before around 50,000 years ago is slight, suggesting that there may have been some event (perhaps a fortuitous mutation) that facilitated humanness – perhaps the acquisition of full language capabilities.

BOX 2.1   **Mind, soul and consciousness**

The ancient Greeks had no doubts about humanness: they distinguished 'body' (the physical part of our existence) from 'soul' or 'spirit'. For them, we were bipartite beings. Plato's dogma of a constant *eidos* separate and independent of physical appearance or behaviour was part of this; it had a deleterious effect on biology for 2000 years, not least because this dualism was absorbed into the Christian understanding of humanness. Aquinas (1225–74) believed the soul was produced by special creation at the moment the embryonic organism was able to receive it. This remains the official Roman Catholic view. In 1996, Pope John Paul II accepting the existence of human evolution ('the human body takes its origin from pre-existent living matter') qualified this by insisting that each body is imbued with a unique soul ('the spiritual soul is immediately created by God').[1]

Much effort has been devoted to seeking the characteristics of this mind or soul. René Descartes (1696–1750) envisaged a 'mind' which controls the body, interacting with the latter through the pineal gland. A US physician went so far as to calculate the average weight of the soul as seven ounces (198 g), based on the weight lost by six men as they died. Fifteen presumed soulless dogs showed no change in weight at death. Such endeavours established nothing. Indeed, from the time of William Harvey (1578–1657) and the discovery of the circulation of the blood (particularly the role of the heart as a pump), the human body has been more and more seen as a machine, with discoverable properties.

In particular, increasing links have been found between the brain and the mind. Aldous Huxley wrote powerfully about the effects of drugs or behavioural conditioning in the 1930s – warning of the implications of outside manipulations in *Brave New World* (some of the more extreme nightmares of determinism he retracted 30 years later in *Brave New World Revisited*, 1958). London taxi drivers faced with the need to learn 320 routes covering 25,000 streets and 20,000 landmarks as part of qualifying for their licence have an enlarged part of the brain which deals with memory. Brain damage may cause profound changes in behaviour or personality; conditions such as schizophrenia or epilepsy can be treated chemically (with drugs). Although recurring claims that particular structures in the brain determine sexual, aggressive or religious behaviour have never been validated, increasingly sophisticated imaging techniques confirm that brain activity correlates rather precisely with conscious human actions and beliefs. The danger is claiming that such beliefs or responses are *nothing but* the result of physico-chemical activity in the brain, but there is no case for maintaining the existence of some separate mind or soul-essence.

Where does consciousness fit? For Francis Crick, 'consciousness depends crucially on thalamic connection with the cortex [i.e. links between different parts of the brain]. It exists only if certain cortical areas have reverberatory circuits (involving cortical layers 4 and 6) that project strongly enough to produce significant reverberations'. But he qualifies this assertion as 'a house of cards. Touch it and it collapses... it has been carpentered together with not enough crucial evidence to support its various parts.'[2] Crick is far from being the only speculator. There has been a spate of 'explanations' of consciousness, such as those provided by authors like the US philosopher Daniel Dennett in his book *Consciousness Explained*,[3] but the truth is that we are a very long way from finding the answer.

The existence of a degree of self-consciousness in some animals has been claimed from the existence of 'mirror neurones' in the brain which lead to an individual's awareness of themselves. But its development in humans remains a mystery. All that can be said is that, with few exceptions, psychologists and neurobiologists concur that consciousness should not be explained by either dualism or the sort of reductionism attempted by Crick. Perhaps the least contentious description of consciousness is that it represents interdependence between mind and brain.

[1] Address to the Pontifical Academy of Sciences, 22 October 1996.
[2] Crick, F.H.C. (1994). *The Astonishing Hypothesis*. New York: Simon & Schuster, p. 252.
[3] Boston, MA: Back Bay Books, 1992.

Can we be more precise about this? It has been said that *Homo sapiens* is not just an improved version of his ancestors, but an entirely unprecedented entity – all of whose intellectual attributes are tied up with language. Many animals have sophisticated means of communicating with each other, but the complexity of language is a uniquely human characteristic. The differentiating feature between ape and human is not the ability simply to make sounds, but the capacity to control those sounds precisely. This was almost certainly a key factor towards humanness. Genetic arguments suggest that it may have been relatively recent, perhaps in the past 40,000 years or even later. Perhaps a significant stage in human differentiation involved the structure of the larynx, tongue and associated structures. The missing ingredient which prevented the chimps developing more complex speech might have been an apparently minor modification of the vocal tract which allowed us finer control and the production of a much greater variety of sounds. In other words, the factor which prevented the chimps developing more complex speech may have been no more than a tiny change in anatomy of the proto-human vocal tract.

BOX 2.2   The supralaryngeal pathway

The supralaryngeal pathway (the route by which air passes from the nose or mouth to the lungs) is entirely different in humans from that in even our closest relatives. In humans, the pathway acts as a sound muffler (Figure 2.3). It cannot do this in apes or monkeys because their tongues are contained entirely in their mouths. Humans develop like apes until around three months of age, then over a period of about nine months the mouth migrates backwards relative to the base of the skull. We do not know for certain when humans diverged in this respect, but it must have been in the early history of the human lineage. More precisely, between 10,000 and 100,000 years ago two nucleotide differences appeared in the human line in a gene (*FOXP2*) which codes for a gene affecting grammar, speech production, non-verbal intelligence and non-speech-related movement of the mouth and face, plus cerebellar development. This gene is highly conserved: mice and primates differ in only one out of the 715 amino acids which make it up. Intriguingly, in one large human family, a single point mutation in the gene (which is on the long arm of chromosome 7, the same chromosomal section involved in the development of some autistic traits) was carried by fifteen out of thirty-one individuals over three generations, and all them had medical problems, including being unable to speak intelligibly. An induced mutation in the comparable (homologous) segment in mice leads to inability to produce ultrasonic sounds and also to cerebellar defects, probably as a result of effects on neuronal migration or of maturation in the cerebellum.

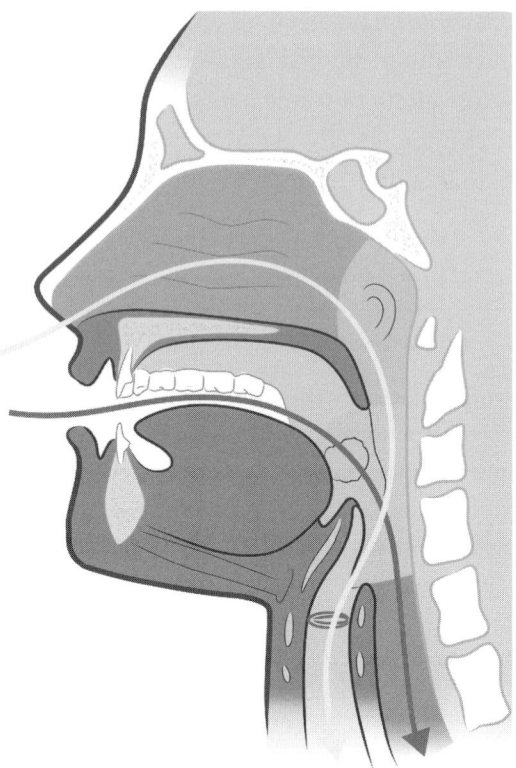

FIGURE 2.3A The supralaryngeal pathway permits the control of the
voice (see Box 2.2). At birth, the tongue is entirely contained in the
mouth, but during the first two years in humans only (not in any
other ape), the roof of the mouth moves backwards relative to the base
of the skull and the tongue begins to move down the neck, carrying
the larynx with it. This makes possible the fine control of speech.
Photo: BSIP/Universal Images Group.

There can be no doubt that the process of 'humanization'
must have depended crucially on sophisticated communication.
This in turn facilitated 'social intelligence', which involves at least
six different faculties: abstract thought, the ability to cooperate in
forward planning, problem solving through behavioural, economic
and technological innovation, 'imagined communities', symbolic
thinking and a 'theory of mind'. A key feature of such a theory of

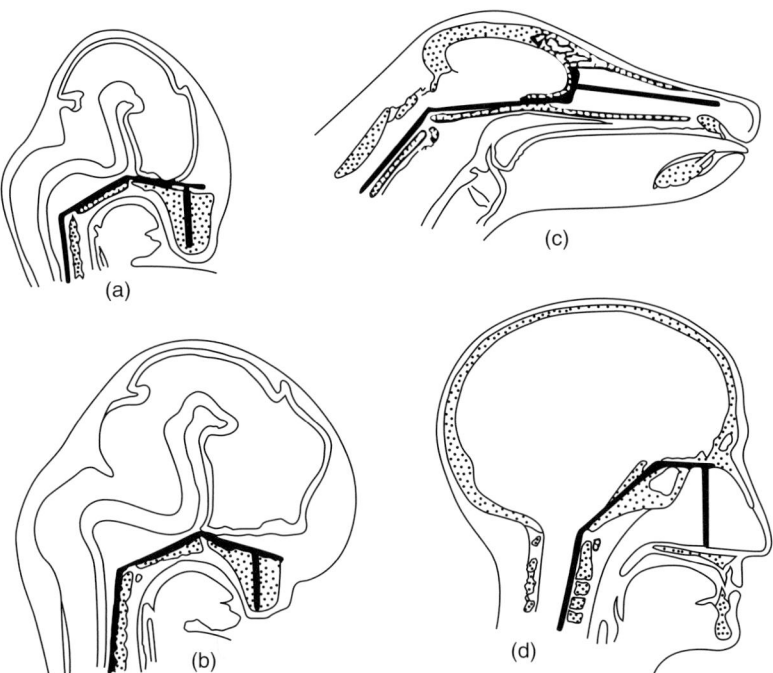

FIGURE 2.3B The heavy line shows the angle the head makes with the rest of the body. In the adult dog (c), the angle is very different from that in the embryo dog (a), but the angle remains unchanged in the adult human (d) from that in the embryonic situation (b).
From Gavin De Beer (1958). *Embryos and Ancestors*, 3rd edn. by permission of Oxford University Press.

mind is the ability to recognize that other individuals may have ideas and desires different from one's own, a faculty obviously helped by the ability to use and understand words. Partial answers to some of these uncertainties blur the behavioural distinctiveness of humans while recognizing that the differences between the apes and us are so great that they may be seen as qualitative. Whatever is finally agreed about the onset of human behaviour, it seems certain that it arose in Africa and was exported from there to the rest of the world.

We cannot be sure about all the steps that led to 'humanness', but there is no doubt whatsoever that the species *Homo sapiens* is a real entity, with a single origin. Clearly there are divisions in humanity – Caucasians, Orientals, Mongoloids, Negroids and so on. At one time it was believed that these represented separate routes to humanness, and that some groups were innately inferior to others. Thomas Jefferson, for example, influentially (and unhappily) argued that Africans are inferior to Whites in intellect and sexual behaviour. The idea promoted by polygenists was that different races had evolved separately on each continent without any recent common ancestor. There is no evidence at all for this: it is important to affirm that the differences which exist in skin colour, academic or athletic prowess, cultural norms, etc. are all secondary and subsidiary to our basic humanness. As far as genetical differences are concerned, there is more variation *within* any 'racial' group than *between* groups.

This does not mean that there are no geographical differences. Clearly there are. Some at least of these result from the effects of natural selection. For example, there are a number of gene variants which seem to protect against malaria (thalassemia, sickle cell); they occur in high frequencies only in areas where malaria is common (or in migrants from such areas – such as Afro-Americans in North America). It has been suggested that many of the differences in blood group frequencies (ABO, rhesus, etc.) reflect differential survival from epidemic diseases in the past (cholera, measles, plague, etc.). An intriguing situation is the ability for adults to digest milk, an ability which is found in most people of northern European descent but at much lower frequencies in other parts of the world. While all healthy children have an enzyme called lactase needed to digest lactose (the sugar found in milk), the enzyme does not persist into adult life in more than half of the world's population. Although cattle herders learnt to turn their milk into cheese or yoghurt where most of the lactose is broken down, they could not digest fresh milk. Around five thousand years ago, at least five mutations for lactase persistence (i.e. the ability for adults to digest milk) appeared. This allowed its possessors to use whole milk throughout life and thus get nutrition and fluid (and survive) even when their crops failed. It was obviously a highly advantageous trait in many situations. The European variant of the gene must have been a factor in the colonization of the more marginal regions of northern Europe. It has spread in the last 4000 years, one of the most rapid changes of gene frequency known.

Life changed radically and became less capricious with the Neolithic revolution, when our forefathers learnt to farm – to herd animals and cultivate useful plants. Analysis of DNA in individuals who lived 8500 or so years ago has shown evidence of selection on genes associated with diet, pigmentation and immunity. Depending wholly on hunting for food is a high-risk strategy: the success rate is often low and can become more difficult if it frightens the animals being hunted. A more efficient way is to manage the hunted animals in some way. As noted above, the first animals to be domesticated were dogs. In eastern and central Europe, hunting must have involved following migrating herds between their winter ranges on the Hungarian plain and the Black Sea coast to their summer grazing in the Jura, southern German highlands and Carpathians. Sites of human habitation have been found along the migration routes and on the edge of reindeer grazing areas. 'Food' was always within reach.

A major event in human history was the apparent abandonment of northern Europe by humans at the height of the last glaciation 25,000 years ago. A denser population developed further south, apparently subsisting on the large reindeer and red deer herds that migrated seasonally through the Dordogne and on salmon and other fish at other times of the year. This was the society which produced the great cave paintings in south-west France and northern Spain. Intriguingly, light (or white) skin and blue eyes only began to appear as our ancestors expanded north into less sunny regions and needed to be able to absorb more sunlight to synthesize vitamin D. Most humans for most of our history have been dark pigmented.

All this brings us back to what actually constitutes humanness. It is obviously more than our physical characteristics and must include our motives and behaviour. A minimum requisite is that we had to attain a level of intellectual development (i.e. brain evolution) before we had any ability to behave ethically. The distinguished geneticist Francisco Ayala has set out three conditions involved in this, all consequent on our biological background rather than any inherent social behaviour or experience:

- Crucially, we had to reach a level of intellectual development which allows us to anticipate the results of our actions. Such a level includes being able to recognize tools as tools rather than merely useful artefacts.
- Secondly, we had to perceive some actions as preferable to others, i.e. to make value judgements, distinguishing 'good' from 'bad'. Ayala sees moral judgements as a particular class of value judgement, involving a regard for others rather than self, allowing altruism to evolve as an evolutionary stable strategy.
- Thirdly, we had to develop the ability to choose between different courses of action.

In other words, value judgements are an integral part of humanness, and inseparable from it. Ayala rejects as unnecessary and without evidence the sociobiological assertion that morality is an evolutionary by-product – an epiphenomenon of our genes.[3] His conclusion may be denied by doctrinaire determinists, but it is not extreme and should not seem contentious to any but extreme reductionists. Karl Popper has commented that 'the fact that science cannot make any pronouncement about ethical principles has been misinterpreted as indicating that there are no such principles, while in fact the search for truth presupposes ethics'.[4]

BOX 2.3   **Sociobiology**

In the *Descent of Man*, Darwin argued that altruism was unlikely to arise from natural selection. He wrote, 'He who was ready to sacrifice his life, as many a savage has been, rather than betray his colleagues,

---

[3] Ayala, F.J. (2006). Biological evolution and human nature. In: Jeeves, M.A. (ed.). *Human Nature*. Edinburgh: Royal Society of Edinburgh, pp. 46–64.
[4] Popper, K. (1977). 'Natural selection and the emergence of mind'. Lecture delivered at Darwin College, Cambridge, 8 November.

would often leave no offspring to inherit his noble nature. The
bravest men, who were always willing to come to the front in war
and who freely risked their lives for others, would on average perish
in larger numbers than other men.'

Half a century later, J.B.S. Haldane qualified this, pointing out that
if there was an unselfishness gene (extending even to the point of
accepting self-sacrifice) and if an individual with such a gene helped
near relatives, then such 'altruistic genes' could spread in families.[1]
He is alleged to have said that he would not try to risk his life to
save a drowning man, although he would be prepared to hazard it for
two brothers or eight cousins. Such relatives would carry some of
his genes and be available to hand them on to the next generation,
even if he perished. In other words, there could be situations where
cooperation (that is, unselfishness) is an advantage to a group of
relatives, even if particular individuals were disadvantaged. Haldane's
argument was formalized by W.D. Hamilton as the concept of
'inclusive fitness'[2] or (as it became known) 'kin selection'; it has
been assimilated into general biology as the mechanism underlying
'sociobiology'.

The 1950s and 1960s saw much interest in biology and behaviour,
shown in popular writings by Konrad Lorenz, Niko Tinbergen, Vero
Wynne-Edwards, Robert Ardrey and Desmond Morris, and expressed
later in television series such as David Attenborough's *Life on Earth*.
The word 'sociobiology' entered common parlance with a 1975
book of that name by the Harvard entomologist, Edward Wilson,
supplemented the next year by Richard Dawkins's *The Selfish
Gene*, a popularization of Hamilton's highly mathematical thesis.
Wilson's book ended with a chapter called 'Man: from Sociobiology
to Sociology', in which Wilson extrapolated conclusions about genes
and behaviour from (mainly) invertebrates to human beings. He
was attacked by both sociologists and socialists, who saw the idea
as contrary to their aims of improving society by manipulating the
environment.

Notwithstanding, sociobiological ideas caught on and stimulated
an immense amount of research. They have also provoked
much dissent, particularly as they apply to mammals (especially

humankind), because of the implications that behavioural choices are programmed (or determined) by genes. Moreover, human social organization extends far beyond family relationships, and most observers believe that the sociobiological mechanisms so far investigated are insufficient to account fully for the widespread acknowledgement of the Golden Rule ('loving one's neighbours as oneself').

Darwin puzzled over this. He wrote in the *Descent of Man*:

> There is no evidence that man was aboriginally endowed with the ennobling belief in the existence of an Omnipotent God. On the contrary there is ample evidence, derived not from hasty travellers, but from men who have long resided with savages, that numerous races have existed, and still exist, who have no idea of one or more gods, and who have no words in their languages to express such an idea. The question is of course wholly distinct from that higher one, whether there exists a Creator and Ruler of the universe, and this has been answered in the affirmative by some of the highest intellects that have ever existed...
> To do good in return for evil, to love your enemy, is a height of morality to which it may be doubted whether the social instincts would, by themselves, have ever led us. It is necessary that these instincts, together with sympathy, should have been highly cultivated and extended by the aid of reason, instruction, and the love or fear of God, before any such golden rule would ever be thought of and obeyed.

[1] *The Causes of Evolution*. Princeton, NJ: Princeton University Press, 1932.
[2] The genetical evolution of social behaviour. I. *Journal of Theoretical Biology*. 7(1): 1–16, 1964.

The way forward seems to be to go beyond formal genetics and recognize two kinds of heredity in higher animals – biological and cultural. Biological inheritance is like that in any other sexually reproducing organism, based on the transmission of genetic information encoded in DNA from one generation to the next by means of the sex cells. Cultural inheritance, on the other hand, is based on transmission of information by a teaching-learning process, which does not depend on biological parentage. Culture is transmitted by instruction and learning, by example and imitation, through books, newspapers and radio, television and motion pictures, through works of art and any other means of communication; it is acquired by every person from parents, relatives and neighbours, and from the whole human environment.

This idea is apparently supported by a major rethink by Edward Wilson.[5] He claimed that kin selection theory is not constructive and has intrinsic defects. He proposed instead the importance of other factors in promoting 'eusociality', which he defined as involving a high degree of social organization, a division of labour, overlapping generations and cooperative care of the young. Wilson argues that eusociality has arisen extremely rarely; he believes it has emerged only about two dozen times in the whole history of life. It depends on the coming together of a 'protected nest' in which to raise young, followed by a lack of immediate dispersal of the young after maturity (which could be the result of a simple genetic change), producing a community with its own characteristics which would be in competition with other communities. Wilson developed this idea from his own expertise with ants, but believes that the origins of humankind could be interpreted similarly. Our own 'protected nest' may have come about relatively recently, but it has been possible because of other changes – as our diet changed from vegetarianism to omnivory, as we learnt to use fire as a social focus and for cooking, and so giving an assured home base for organizing and cooperating in hunting, thus providing for those in the 'nest'. Such a community will involve an extended family, but not one limited or restricted to close relatives.

The argument is that cultural adaptation has largely taken over from biological adaptation in humankind because it is both a more

---

[5] Wilson's change of mind and rejection of the parts of sociobiology dependent upon kin selection provoked an article critical of his new thinking signed by 137 evolutionary biologists. Wilson's reply was that Einstein's theory of relativity had given rise to a paper in 1921 criticizing his General Theory of Relativity (at a time when it was effectively proved), with signatures from 100 physicists.

rapid mode of adaptation and because it can be directed.[6] A favourable genetic mutation newly arisen in an individual can spread to a sizeable part of a species only through a large number of generations. In contrast, a new scientific discovery or technical achievement can in principle reach the whole of humanity in less than one generation. Moreover, whenever a need arises, culture can directly pursue the appropriate changes to meet the challenge. This differs radically from biological adaptation, which depends on the fortuitous availability of a favourable mutation (or a combination of several mutations) at the time and place where the need arises.

Put crudely, although we are DNA replication machines, we are much more than that; we are both a part of and apart from the natural world in which we live.

BOX 2.4 **René Dubos**

René Dubos (1901–82) was born in a village north of Paris. He trained as an agriculturist and became interested in soil microbiology, moving to the USA in 1924 where he worked for most of his life at the Rockefeller Institute in New York. His research convinced him of the complexity and interactions of soil bacteria, which had been largely ignored before his time because of the technical difficulty of culturing them in isolation. He was what is often called a 'hard' scientist. He discovered antibiotics produced by soil bacteria, but his influence spread much wider from the microbiological community because of the way he extended his thought into the implications of human thought and action. Towards the end of his life he wrote:

> Our separation from the natural world leaves us with a subconscious feeling that we must retain some contact with wilderness and with as

---

[6] The idea that humans have passed from a phase of biological evolution to one of cultural evolution is not new. Julian Huxley called the latter phase 'psycho-social evolution'; C.H. Waddington called it 'socio-genetic evolution'. Darwin's contempory, A.R. Wallace, described the application of natural selection to tribes with advantageous 'mental and moral qualities' compared to their neighbouring tribes, in 1864.

wide a range of living things as possible... Correcting the damage done
to nature by industrialisation is probably well within our powers, but
to formulate new positive values for modern life will be much more
difficult... The futures we invent are viable only if they are compatible
with the constraints imposed by the evolutionary past. This does not
mean that the desirable future is one which would take us back to the
pre-technological womb. But it does mean that the unchangeable laws
governing human nature and external nature must be kept in mind
whenever plans are made to change the conditions of life.[1]

Dubos popularized (and probably invented) the slogan 'think
globally, act locally'.[2] He was dismissive of calls common among
environmentalists to look back to St. Francis of Assisi as a model. He
regarded St. Benedict, with his emphasis on management, as a more
worthy example.

Maurice Young, the Secretary-General of the United Nations
(UN) Conference on the Human Environment in Stockholm in
1972 (p. 188) – the first major international conference on the
environment – was so impressed with Dubos's approach that he
invited him to join the economist Barbara Ward in preparing the
background document for the conference. Dubos persuaded Ward
that she had become too focused on problems and emergencies;
she needed to take on board how humans must take responsibility
for their decisions. They came to a common mind and together
defined the fundamental task of humankind as 'to devise patterns
of collective behaviour compatible with the continued flowering
of civilisations'.[3] This became the concept of 'development
without destruction' in the statement produced at the end of the
Stockholm Conference; it developed into the notion of 'sustainable
development', which features in virtually all subsequent
discussions.

[1] Dubos, R. (1973). *A God Within*. London: Angus & Robertson,
pp. 166, 280.
[2] The extraordinary Patrick Geddes (Professor of Botany in Dundee, then
Professor of Sociology in Bombay), who inspired three of the four original
members of the British Vegetation Committee, fore-runner of the British
Ecological Society (p. 128), developed a similar concept in 1915 in terms of

town planning. Dubos's more general use of the term certainly pre-dates that of Friends of the Earth founder, David Brower, to whom it is often attributed.

[3] *Only One Earth: The Care and Maintenance of a Small Planet.* New York: Norton, 1972.

We are certainly moulded by our surroundings, both historically and genetically, and by the circumstances of our own lives, but we are also conscious agents. It would be stupid to assume that we can treat our environment as nothing more than a convenience or décor. At the height of the tumult of industrialization in the mid-nineteenth century, John Ruskin wrote:

> God has lent us this earth for our life; it is a great entail. It belongs as much to those who come after us and whose names are already written in the book of creation, as to us; and we have no right by anything we do or neglect to do to involve them in unnecessary penalties, or deprive them of benefits which it was in our power to bequeath.[7]

Others have argued similarly. 'A true conservationist is a man who knows that the world is not given by his fathers but borrowed from his children' (often attributed to John James Audubon); 'I recognize the right and duty of this generation to develop and use our natural resources, but I do not recognize the right to waste them or to rob by wasteful use the generations that come after us' (Theodore Roosevelt); [8] 'The ultimate test of a moral society is the kind of world

---

[7] The Lamp of Memory. In: *The Seven Lamps of Architecture, Volume 1*. London: Smith, Elder and Co, 1849.

[8] Roosevelt, T. Speech given at Osawatomie, Kansas, 31 August 1910. The Native American Chief Seattle is sometimes claimed by environmentalists to have propounded a 'Fifth Gospel' in a speech in 1854: 'How can you buy or sell the sky, the warmth of the land? The idea is strange to us. If we do not own the freshness of the air and the sparkle of the water, how can you buy them... This we know – the Earth does not belong to man – man belongs to the Earth.' Chief Seattle may well have valued the environment, but these words were actually spoken a hundred years later in a film called *Home* produced in 1972.

that it leaves to its children' (often attributed to Dietrich Bonhoeffer); and 'The future starts today, not tomorrow' (often attributed to Pope John Paul II).

The British Government's submission to the UN Conference on Environment and Development in Rio in 1992 (The 'Earth Summit') repeated the same analogy: 'Mankind has always been capable of great good and great evil. That is certainly true of our role as custodians of the planet... It was the Prime Minister [Margaret Thatcher] who reminded us that we do not hold a freehold on our world, but only a full repairing lease. We have a moral duty to look after our planet and to hand in on in good order to future generations.'[9] This idea of 'moral duty' involves the characteristics of humanness as envisaged by Ayala. This is the relevance of who we are – our humanness; it is where humanness comes face to face with our environmental attitudes.

The Neolithic was the time when we moved from being buffeted by our environment to some ability to take charge. It was marked by attempting to manage wild food resources. A key step would have been sowing previously harvested grain. This meant a more settled lifestyle, which in turn allowed specialization of occupation and also a need to establish systems of land tenure. Living in a settled community exposed our forebears to increased danger from infectious diseases and intensified social conflicts. It marked the development of a much more complicated attitude to the environment than that of the earlier hunter-gatherer phase of existence. A small hunting band could respond to changed conditions by moving to new hunting grounds. A farming village could receive emergency support from neighbours. But as villages became towns and then cities, they were unable to adjust easily. Having grown as a successful defence against small catastrophes, cities became increasingly vulnerable to larger ones.

[9] *This Common Inheritance: A Summary of the White Paper on the Environment.* London: HM Stationery Office, 1990.

Humankind necessarily developed different strategies for dealing with environmental pressures in different parts of the world. Are there any generalities? Can we discern any patterns in how our forefathers adapted and survived? There is more than academic curiosity in this, because if they had not survived, we would not be here now.

FURTHER READING

Alexander, D. (2011). *The Language of Genetics.* Conshohocken, PA: Templeton Press.

Ayala, F. (2009). Being human after Darwin. In Northcott, M.S. and Berry, R.J. (eds.). *Theology After Darwin.* Milton Keynes: Paternoster, pp. 89–105.

Darwin, C. (1871). *Descent of Man.* London: John Murray.

Diamond, J. (1991). *The Third Chimpanzee.* New York: Barnes & Noble.

Jeeves, M.A. (ed.) (2004). *From Cells to Souls and Beyond.* Grand Rapids, MI: Eerdmans.

Shugart, H.H. (2014). *Foundations of the Earth.* New York: Columbia University Press.

Stringer, C. (2011). *The Origin of Our Species.* London: Allen Lane.

Wagner, R. and Briggs, A. (2016). *The Penultimate Curiosity.* Oxford: Oxford University Press.

Wilson, E.O. (2014). *The Meaning of Human Existence.* New York: Liveright.

3 Striving with Nature

The Neolithic revolution (a term devised by the British archaeologist Gordon Childe in 1923) was a major step towards greater environmental emancipation. It was more evolution than revolution, a response to the perceived advantages of agriculture rather than an inevitable result of cultural change. It opened the floodgates of change, leading to the existing small bands of hunter-gatherers joining into non-nomadic societies living together in communities. Evolutionary change does not happen for its own sake, but results from altered pressures which force modifications in those under pressure. Modern hunter-gathers seem to be able to sustain themselves with only a modest amount of work; they even have a fair amount of leisure time. Agriculture must have had a significant advantage over hunter-gathering because it apparently spread rapidly (Figure 3.1). It may be that a sedentary existence allows a wider range of foods to be harvested than by hunting, leading to an increased confidence of food available locally. It certainly makes storage more feasible. The development of grinding techniques meant that more food could be extracted from grain, but such equipment would further constrain a group's mobility.

Around 10,000 years ago, virtually all populations existed as small hunting groups; by 2000 years ago the overwhelming number of people lived by farming along river valleys. The immediate value of crops and livestock was as food. However, they also provided fibres for clothes, blankets, nets and ropes. The bones of domestic animals were used to make artefacts for hooks or personal adornment before metals were worked. Animal hides could become containers for liquid. Larger animals were on hand to be used as transport, both for humans and hauling goods. Wheeled vehicles began to appear from

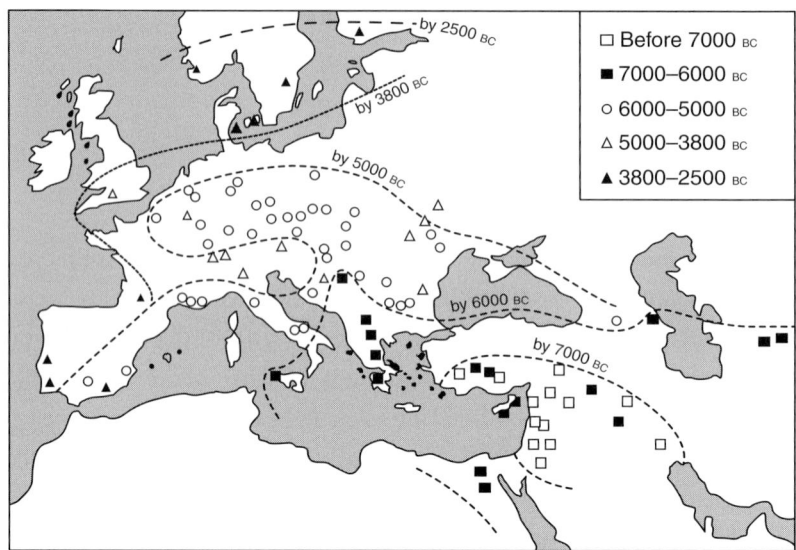

FIGURE 3.1 The rate of spread across western Eurasia of the key food crops which were domesticated from species growing in the 'fertile crescent' of the eastern Mediterranean.
From Daniel Zohary and Maria Hopf (2001). *Domestication of Plants in the Old World*, 3rd edn. by permission of Oxford University Press.

about 6000 BC. All this introduced a greater security into existence. It also marked the time when we began to change from being a pawn in our environment to becoming a partner with it.

Human numbers exploded in the Neolithic. There does not seem to have been any major change in individual life expectancy, so the growth must have been due to increases in birth rates and neonatal survival. The interval between births in modern hunter-gathering groups is about four years; in rural populations in Europe two centuries ago, it was two and a half years. At the time settled agriculture began around 10,000 years ago, the world population was around fifteen million individuals; by early Christian times there were something like fifty million people in the Roman Empire alone. It is estimated that there were a mere 2000 people in the British Isles 12,000 years ago at the time the islands were cut off from the main-land of Europe, all of them hunter-gatherers leaving few traces of their presence and having little impact on the land or its creatures. By 5000 years ago the population had grown tenfold to around 20,000, and was continuing to increase rapidly.

Community gave security. More food was often grown than was needed for immediate use. This meant that food shortages became less common and often more predictable. Surpluses gave farmers a clear advantage over hunters, because they could store their surplus and tide over unfavourable conditions. They also made it possible for non-food-growing specialists to appear, in particular administrators and rulers. This led to strata within society, with the emergence of elites able to dominate their communities and monopolize decision making; the old equalitarian hunter-gathering societies split; chiefs and serfs, lords and villeins, oppression and slavery emerged.

The situation in England changed further with the Norman conquest. The Normans had little interest in farming, but they revo-lutionized the laws of land tenure. In order to establish their position, they stabilized and perfected the feudal system – and, through their love of hunting and the establishment of Royal Hunting grounds, drew up a web of strict Forest Laws. At the end of the twelfth cen-tury something like a third of England was subject to Forest Law, restricting wood-cutting and grazing by common people. Meanwhile, the parish system evolved experientially. The countryside changed as communities turned their back on the network of roads and towns established by the Romans. The basic need of a village was a water supply, which formed an essential nucleus for every community. The settlement pattern was pretty well established by the time of the Domesday survey in 1086. There were clear regional differences. The least-farmed areas were the heath and breckland areas of the Weald, New Forest, Dorset and Surrey, and large parts of northern Britain. By the fourteenth century the dominant arrangement involved open field farming (Figure 3.2), with strips allocated to individuals within two or three open fields and a rotation of crops (winter wheat/rye/ spring barley/oats/legumes/fallow). There was a need for a balance between arable, pasture and woodland to provide

FIGURE 3.2 Field networks in the Cotswolds, Gloucestershire, England –
a pattern characteristic since early medieval times.
Photo © fotoVoyager.

vegetables for food, animal food and timber. A key limiting factor was
sufficient feed for overwintering animals. Such a parish unit would
have been largely self-sufficient with its own specialist craftsmen
(wheelwright, carpenter, smith) and capable of being able of providing
tithes to support its church and priest. Proper functioning depended
on cooperation. A cultivator would have to observe the husbandry
practices and calendar of the community. It would have been natural for the village of early settlement days to develop a leader, to
be replaced in time by a hereditary headman or Lord of the Manor.

Towns grew up as sites for markets. As Christianity spread, so did the increasingly influential monastic communities.

In Scotland, the unit was more often a farm rather than a village. A quarter or so of a farm's land might be intensively cultivated and manured 'infield', with shifting cultivation and rough grazing practised in the 'outfield'. The earliest settlements must have been in easily cultivable areas, but as they grew in size and numbers, they would have had to spread into less welcoming habitats. This meant the newly settled farming communities had to modify their environment – often radically – by draining swamps and clearing forest.

These same challenges faced the early European colonists to North America, where their travails are well documented. The newcomers found a continent that was immense, seemingly inexhaustible and certainly dangerous. It required hard work, technical skill and luck to succeed in the wilderness. The early settlers followed their European lifestyles and social organization, especially in New England, in the Virginia tide-water areas, and in areas settled by Germans from Rhineland-Palatinate. Then a new individual appeared – the frontiersman or pioneer, following the retrenchment after Independence. The frontiersman became the mythical hero who opened new lands to settlement through his courage, skill and derring-do. Together with his surrogates – the mountainmen, the cowboys, the loggers, the flatboat men – they came to represent charismatic characteristics which Americans looked up to as valuable and desirable. But wherever in the world farming spread – and similar frontiers existed in many places – there were problems of climate and soil fertility and the ever-present hazard of overgrazing, with the danger of increasing deserts, made worse in marginal areas by the browsing habit of goats.

All this involved new habits, new techniques, new social structures and new problems. As with all innovations, there must have been all sorts of failures and recurrent disasters, perhaps most acutely from disease – able to establish and spread more easily within the new communities than in itinerant hunting bands. Some diseases would have been known in pre-Neolithic times. Hookworms would have affected fishers; parasites like trichinosis or anisakiasis would have infected humans from raw or partly cooked meat; or, more grimly, the prion disease kuru (related to Creutzfeldt-Jakob or 'mad cow' disease), transmitted by cannibalism. But always there were (and are) animal diseases which 'jump' to humans who have no innate resistance to them, and which are often symptomless in their animal host. It is likely that measles, smallpox, influenza and tuberculosis all came from animals. Much more recently, AIDS originated from a virus carried by chimpanzees in the Democratic Republic of Congo. Such diseases ('zoonoses') may have afflicted hunter-gatherers, but the small size and isolation of their groups would have meant that they did not spread. The growing Neolithic villages and towns gave the new diseases opportunities to attack much larger groups, and perhaps spread between them. Populations usually build up resistance to 'new' diseases, but this takes time and original infections may be devastating. The earliest documented case is an epidemic in the eleventh century BC recorded in the Hebrew Bible where the Philistines of Ashdod are described as being afflicted with plague, probably transmitted by rodent-borne fleas.[1] Thucydides describes a typhoid epidemic in 430 BC which apparently began in Ethiopia, spread through Egypt and Libya, and then reached Athens where a third of the population died.

However, a series of events (bad weather, failed harvests and, most catastrophically, the Black Death) disrupted all this. The Black Death apparently started in China around 1338, killing around half the population there, followed by a third of that of the whole world. It reached Britain in 1348. Between a third and a half of the British population died. Labour became scarce, so that individuals were able to bargain for their pay. Many large estates were broken into smaller units, with a richer peasant class winning out over a landless

[1] 1 Samuel chapter 5, verse 6.

labouring group. There was a swing from labour-intensive arable farming to pasture, particularly of sheep keeping. A class of wealthy yeoman farmers emerged.

The plague repeatedly re-emerged in later years, the last outbreak in Britain being the Great Plague of 1665. The effect of 'European diseases' on the native populations of the Americas and the Pacific global scene was catastrophic. Smallpox reached Mexico in 1520 when an infected slave arrived from Spanish Cuba. Something like half the Aztecs succumbed, and their army was demoralized because the invading Spaniards seemed immune to the disease. Pizarro landed in Peru in 1531 with an army of a mere 168 men, but conquered because a high proportion of the native Inca population had died from smallpox. The threat of such powerful pestilences must always have hung over communities, particularly while the causes of the disease remained unknown. Recent experiences with diseases like Ebola show that this is a continuing hazard, even when backed up with modern knowledge about causes.

BOX 3.1   Cholera

Cholera originated in India, becoming widespread throughout the world in the nineteenth century. It has killed tens of millions since then. It reached England in 1831 from an infected sailor in Sunderland. Two months later it had spread to London. Around 52,000 people died. Another outbreak in 1848 killed even more. It recurred again in 1854. In the Soho district of central London 127 people succumbed over a three-day period; ten days later, 500 people were dead. A London doctor, John Snow, doubted the conventional wisdom of the time that cholera (and other diseases, such as plague) were caused by polluted air or 'miasma'. He argued that the disease entered the body through the mouth. He plotted the homes of affected people on a map, and found they clumped around an area in Broad Street (now Broadwick Street) (Figure 3.3). Only ten deaths occurred in houses near another street pump – and five of those

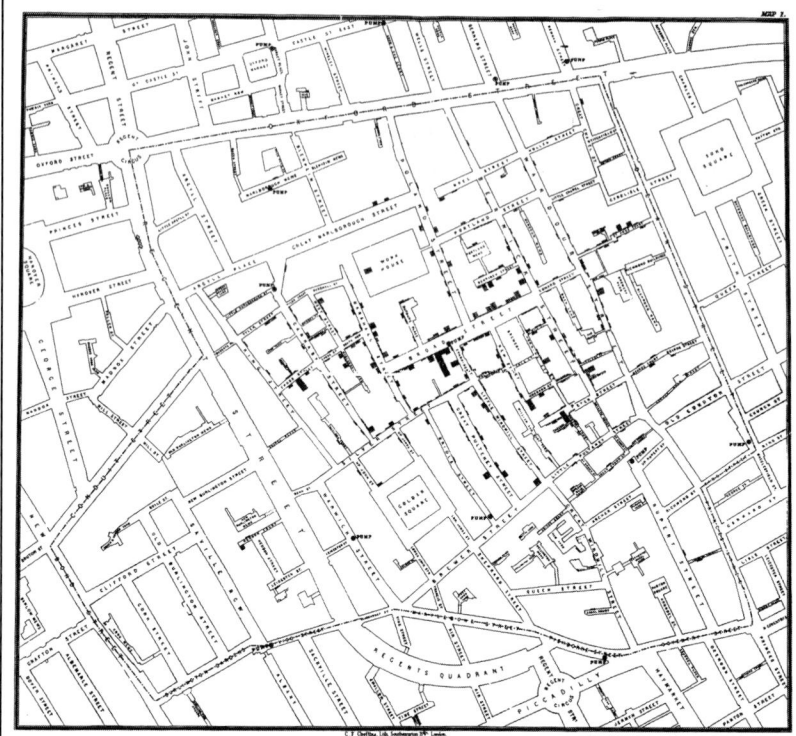

FIGURE 3.3 Map of Soho (central London) prepared by John Snow in 1854 showing the houses of people with cholera. On the basis of their distribution, Snow was persuaded that their source centred around a particular communal pump. Removing the handle of the pump stopped the epidemic.

regularly drew water from a pump in Broad Street. He removed the handle of the Broad Street pump, cutting off the local water supply; cases of cholera immediately began to diminish. It was later discovered that the Broad Street well had been dug less than a metre away from an old and leaky cesspit. Like any challenge to accepted ideas, objections were raised to Snow's theory. One was that none of the monks in a monastery next to the well were affected with cholera. The answer proved to be simple: the monks drank only beer, which they brewed themselves.

The irony is that a committee established to examine the causes of the 1854 epidemic found cholera vibrios (bacteria) in water and on the soiled clothes of victims, but rejected their involvement in the disease on the grounds that they were 'a product of enteric decomposition and they were found also in samples from those who had died of different diseases'. The germ theory of disease only became accepted through the work of Louis Pasteur in the 1860s; the cholera vibrio itself was described by Robert Koch in the 1880s.

Superimposed on the devastation of disease were uncertainties about the land itself. Coastal communities are always at risk from flooding. Around 25,000 years ago a stretch of land between 2 km and 40 km wide and 600 km long was lost in the eastern Mediterranean. About 8000 years ago an area as big as Iceland broke away from the edge of the Norwegian continental shelf, causing a 20-metre-high tsunami which swept over the Scottish coast, even submerging some of the Shetland Islands. Then, 4000 years later, the Neolithic village of Skara Brae in Orkney, an icon of the period in northern Europe, was deserted by its inhabitants, perhaps due to a combination of coastal erosion and falling productivity of the land close by; it was subsequently overwhelmed by sand. It is obviously difficult to put ourselves in the shoes of the earliest farmers, but what about current ones? In 2005 Hurricane Katrina led to 1800 deaths in the southern USA, displaced 1.2 million people and caused $110 billion in damage. Seven years later, Superstorm Sandy cost the state of New York $32 billion. The current rise in sea level as a result of global warming is already causing problems for low-lying islands in the Indian and Pacific Oceans. In Bangladesh ten million people are at risk on land lower than one metre above high tide level.

Other effects of climate may cause problems, affecting subsistence farmers more than city dwellers. Those living in tropical latitudes would not have had to cope with much change in day length, although they might have had to recognize wet and dry periods, and perhaps seasonal availability of their animal prey. Our earliest ancestors who left Africa and moved north would have encountered more seasonal fluctuation and probably greater climatic uncertainty than in their homeland. It would have been essential for them to understand the regularity of the seasons; the winter and summer solstices would have been key elements in their life. We can extrapolate something of their existence from studies of contemporary uncontacted tribes, but it is far from our modern way of life, wherever we are. About 6000 years ago, the Indian monsoon track shifted southward and the rainfall pattern changed. Farmers in the Middle East would have had to depend more on water from rivers and had to develop a system of canals for irrigation. Around 1000 years later, southern Mesopotamia was a mosaic of competitive city states, centred around jealously guarded irrigation canals. Then in 2200 BC a volcanic eruption to the north produced ash which veiled the Sun and coincided with three centuries of drought. The once-fertile plains were turned into near desert. One of the greatest centres of early civilization was the city of Ur, founded 6000 years ago; its site is now a sun-baked desert.

BOX 3.2 **Volcanic eruptions**

Volcanic eruptions were particularly terrifying because they might occur far from affected areas and thus their damage might have no obvious cause. The disorders following the fall of the Roman Empire at the end of the fifth century were followed in AD 536 by a year when (in the words of Bishop Zacharias of Mytilene in *The Syriac Chronicle*), 'the sun began to be darkened by day and the moon by night, while the ocean was tumultuous with spray... From the large and unwonted quantity of snow the birds perished... there was distress among men from the evil things.' It marked the beginning of a decade of cold and turmoil from Italy to Iceland. A quarter of the population of the Byzantine Empire died of plague: 10,000 deaths a day at its peak. Prophecies of the end of the world flourished. In China, the Northern Wei dynasty collapsed. In retrospect it seems fairly certain that all these effects stemmed from the eruption of Mount Ilopango in San Salvador, releasing around 84 km^3 of pumice and ash into the atmosphere. The Maya civilization a few hundred kilometres north survived, but retreated from the highlands. It also changed; no elaborate monuments (stelae) were produced after this time.

An even bigger eruption occurred in 1815, leading to 'the year without a summer' in eastern North America and western Europe the following year. It was a time when Europe was beginning to recover from the Napoleonic wars, but was suffering social unrest because of the need to adjustment to an unaccustomed peace. Unknown to those in the western hemisphere, 'the year without a summer' was the consequence of a massive eruption of Mount Tambora on the far-distant Indonesian island of Sambawa the previous year. It ejected around 150 km^3 of material into the atmosphere (almost twice as much as from Mount Ilopango and about a hundred times greater than that from Mount St Helens in 1980, and a thousand times that of the volcano in Iceland in 2010 which seriously disrupted commercial flying). Orange-coloured snow fell in Hungary in June 1816. More disastrously, as we know from more recent studies, sulphur must have accumulated in the atmosphere as sulphuric acid aerosols, which would have persisted for five or six years, forming a veil and filtering out sunlight and thus cooling the Earth's surface three or four degrees. There was a failure of harvests over a wide area. The rice crop in Yunan Province (south-west China) was wiped out three years running. Sea-borne trade cushioned the consequences for Britain but massive amounts of grain had to be imported and the price of bread doubled. The problems in landlocked central Europe were greater. Byron and Shelley, escaping the British winter in Switzerland, chronicled the storms and flooding around Geneva in detail. In London, *The Times* thundered about God's wrath; church-going increased significantly. There was chronic famine, social unrest and a surge of radicalism which persisted for several years, culminating in the 'massacre' of Peterloo in 1819 (p. 127). But the physics of the atmosphere which caused the changes in weather only began to become clearer in the middle of the nineteenth century through the work of Ferris, Galton and others.[1]

[1] Moore, P. (2015). *The Weather Experiment*. London: Chatto & Windus.

It is sometimes claimed that people in northern climes have an advantage because their climate is more conducive to activity and initiative than those living in more torrid tropical or sub-tropical latitudes. There is no real evidence for such a conclusion. Notwithstanding, life in a marginal environment is intrinsically challenging, often forcing experiment and inventiveness. Archaeologists frequently record upland farms deserted as the occupants were forced to retreat to more assured sources of food at the coast. However, the most important developments were those which allowed climatic changes to be ignored – technological improvements in the widest sense. The Spanish conquistadors triumphed in South America despite their relatively small numbers because they had horses, allowing rapid deployment of troops; they had steel weapons and armour, giving them much better personal protection than their enemy; and the Native Americans were much reduced in number by epidemic smallpox, introduced from Europe.

The northern spread of agriculture can be traced by the pollen in sediments: the proportion of tree pollen falls as woodland is cleared, and the proportion of grass and cereal pollen increases. The climate is never constant, but at times it has been particularly unstable. There was a warmer period ('Holocene Climate Optimum') at the beginning of our modern era, lasting from 9000 to 5000 years ago, which no doubt encouraged population expansion into new areas. The earliest known settlements were small hamlets in southern Mesopotamia in 5800 BC. Some of these grew rapidly over the next millennium.

A period of slow cooling followed the Holocene Optimum, with sharp drops in AD 542 and AD 536, perhaps due to dust in the atmosphere following a volcanic eruption. There were crop failures as far apart as Ireland, China and Peru, and some historically recorded events may be the result – the burying of treasure by Scandinavian elites, the decline of the huge city of Teotihuacan in what is now Mexico, even the rise of Islam. Then there was a 'Medieval Warm Period' from 950 to 1250 when Greenland and probably North America were colonized from Scandinavia, followed by another cold period, the 'Little Ice Age', lasting until the mid-nineteenth century. Glaciers in the Alps advanced significantly in the mid-seventeenth century, crushing entire villages. Similar expansions took place in various parts of the world, although the timing of these is not the same in different parts, suggesting that they were independent events.

During the Little Ice Age, Dutch canals and the River Thames in London froze. A frost fair was held on the Thames in 1608 and periodically thereafter. In 1683–4, the Thames was frozen for two months. The last frost fair took place in 1814. In 1780, New York harbour froze, allowing people to walk from Manhattan to Staten Island. There were famines in France in 1693–4, Norway in 1695–6 and Sweden in 1696–7, leading to a 10 per cent reduction in population numbers. Villagers in the Alps were reduced to using flour made from ground-up shells of nuts mixed with barley and oats for bread making. The winter of 1794–5 was particularly harsh. The Dutch fleet was frozen in its harbour at Den Helder; the invading French army used the frozen rivers as highways. The Norse Greenland settlements shrank and then disappeared around 1500.

Despite the number and variety of uncertainties – from climate, disease and the ever-changing political situation – in Britain at least an understanding of the environment came to mean incorporating, or at least tolerating, these rogue elements. Agriculture moved from acceptance of outside influences to 'improvement' and a reduced dependence on circumstances. The Royal Society, founded in 1660, set up a committee in its first year to explore the potential of the potato, which had been introduced into Europe in the late sixteenth century and proved easy to grow and less susceptible to spoilage and bad weather than cereal crops. It could be cultivated in small plots and seemed an ideal crop for workers moving into urban areas.

The breaking up of the feudal system following the Black Death resulted in many yeoman farmers buying land and fencing it so that they could make innovations, as well as contain their livestock. Such enclosures were unpopular from the start, depriving agricultural workers of their stake in the land. New crops were introduced to improve fertility. The old open fields were dominated by grain production; the enclosures encouraged the spread of potatoes, carrots and turnips.

The eighteenth century was a time of taming, order and structure. New varieties of barley and wheat meant that crops became stronger and less weedy. New rotations brought in rape and flax. Field drains meant heavy soils became more manageable. In 1701 Jethro Tull invented a mechanical drill for planting and weeding, accelerating the use of root crops. Turnips and clover were introduced as rotated crops, replacing the tradition of leaving land fallow every second or third year. Field enclosures had begun by the twelfth century, but the development of crop rotation speeded them up. A fodder crop clearly had to be protected against grazing. Tull had a disciple in Charles, Second Viscount Townshend ('Turnip Townshend'), who revolutionized faming in East Anglia by instituting a four-year crop rotation. He divided his fields up into four – wheat in the first field, clover (or ryegrass) in the second, oats or barley in the third and turnips or swedes in the fourth. The turnips were used as fodder to feed livestock in winter. The problem of winter feed became much less acute and there was a possibility of stock improvement. Clover and ryegrass were grazed by livestock. Using this system, he found that he could grow more crops and get a better yield from his land.

It would be wrong to regard this period of 'improvements' as a time of universal improvement in the lot of agricultural workers. There was chronic overpopulation and repeated evictions and emigration. In his *Rural Rides*, William Cobbett (1763–1835) described well-woolled sheep and contented cows, but railed against the poverty of many farm workers living in hovels and subsisting on a diet of potatoes and tea. John Clare (1793–1864) railed similarly in verse. In the 1830s the Swing rioters reflected rural labourers' anger at being replaced by machines. Pressure on land meant increasing reclamation of poorer soils and fen lands. On the positive side, it was a time when farm surpluses appeared, particularly in the production of wool, which brought great wealth (and beautiful churches) to some, although consolidation of farm holdings into bigger units consigned many others to hard labour and bare subsistence. Parliament passed a General Enclosure Act in 1801 to control ad hoc local enclosures in favour of a more systematic approach. More than 3000 Enclosure Acts were enacted in the nineteenth century.

This evolving rural symbiosis was shaken – shattered might be a better word – by industrialization and its complement, the urbanization necessary to run the increasingly complex factories which housed the machines. The annual rhythm of human–environment interaction was overridden by co-opting what might be called artificial energy sources. Water and wind power had long been used as partners in human endeavours. Industrialization involved harnessing energy from coal-fuelled and steam-driven machines. Town dwellers lost their dependence on their environment. It became a problem to be overcome rather than a partner to be nurtured and accommodated, while the new ways of life introduced a new range of challenges – transport of foodstuffs, disposal of wastes, disorders of urban living (deficiency diseases, overcrowding, economic and social exploitation). A new set of skills and controls for planning and construction became necessary. Correcting serendipitous mistakes became ever more important. The Great Fire of London gave the opportunity (only partly fulfilled) to plan a better city for Londoners, but in general improvements became increasingly difficult as populations increased and infrastructures became more fixed.

Our Neolithic ancestors were pawns in a hostile environment. Their descendants learnt to partner and cooperate with their surroundings. Modernity is an attempt to overrule and ignore the environment by insulating people from it – which make hurricanes, floods, earthquakes and extreme weather almost insulting.

FURTHER READING

Diamond, J. (1997). *Guns, Germs and Steel*. London: Jonathan Cape.

Fagan, B. (2004). *The Long Summer*. New York: Basic Books.

Glantz, M.H. (ed.) (1994). *Drought Follows the Plow: Cultivating Marginal Areas*. Cambridge: Cambridge University Press.

Hoskins, W.G. (1955; revised edition 1988). *The Making of the English Landscape*. London: Hodder & Stoughton.

Kington, T. (2010). *Climate and Weather*. London: Collins.

Rackham, O. (1986). *The History of the Countryside*. London: Dent.

Reynolds, F. (2016). *The Fight for Beauty*. London: Oneworld.

Salaman, R. (revised edition 1985). *The History and Social Influence of the Potato*. Cambridge: Cambridge University Press.

Ucko, P.J. and Dimbleby, G.W. (eds.) (1969). *The Domestication and Exploitation of Plants and Animals*. London: Gerald Duckworth.

4 Nature's Study

It is possible to read the human story as a narrative of ever-increasing attempts (and some success) to become less and less dependent on the whims of the environment – those physical, biological and social discomforts. It certainly seems a more accurate description of our progression than that other narrative – that our existence is directed towards little more than spreading our selfish genes. Leaving aside for the moment the question of how much this hard-earned freedom can be regarded as a positive, it is legitimate to enquire about some of the nudges that have enabled it. Intriguingly, a few influential people and their ideas have significantly fuelled our progress. Jethro Tull and Turnip Townshend, who introduced the mechanization of agriculture, were two of these. Perhaps their contemporary, the land-scape architect 'Capability' Brown (1716–83), should be added. He introduced a fashion for wildness in something like 260 of the great estates of Britain and thereby changed the landscape of large tracts of the nation. His work influenced an awareness of 'natural' nature in contrast to the artificial formality of previous centuries that show-cased and reflected our attempts to 'tame nature'.

A key moulder of attitudes in the next generation was John Ruskin (1819–1900). He is probably best known as a painter and art critic, but his influence and interests were much wider. He first came to prominence in defending the paintings of J.M.W. Turner (who emphasized realism in his landscape paintings) and then for his support of the Pre-Raphaelites. His relevance here is his insis-tence, following Capability Brown, that the principal role of an artist is 'truth to nature'. He was passionately fond of the countryside and fought for social justice as a counter to the single-minded pur-suit of wealth. He hated his contemporaries' obsession with money, machinery and mechanization, which he regarded as driving out the beautiful. He took up Wordsworth's defence of the Lake District, particularly reviling the spreading railways of his time (p. 8). He had a utopian vision of a world where social justice, human effort and the intrinsic beauty of art and architecture would join in creating a society in harmony with itself. He inspired William Morris and the founders of the National Trust (p. 154), as well as being admired by others as disparate as Mahatma Gandhi and the founders of the Labour Party.

Ruskin was highly important in stimulating interest in the environment. But he was far from being the first or only prophet. There were other significant movers and shakers who have allowed us to dissect and – to some extent at least – understand our environment. The polymath Patrick Geddes (p. 43) speculated that the 'botanical geographer' might be the 'twin brother to the landscape-painter'.[1]

We have already referred to the 'tipping point' of 1543 produced by the astronomer Copernicus and the anatomist Vesalius (Box 1.4), but they were standing on the shoulders of giants, albeit that some of their predecessors gave very dubious support. There were certainly giants in the ancient world. A significant progenitor has to be Aristotle (384–22 BC), who moved protoscience towards science by recognizing that a true understanding of the world required actual observations and their organization; speculation and tradition were not enough. Charles Darwin wrote, 'Linnaeus and Cuvier have been my two gods... but they were mere schoolboys to old Aristotle.'[2] Aristotle was a pioneer not only in collecting observations but also in synthesizing and reconciling them. He is best known for his later works on physics and philosophy, but his methodology and intellectual discipline came from two years studying biology on the island of Lesbos. Scientific study (although it was not called that in those days) preceded his massive contributions to social and metaphysical

---

[1] *Chapters in Modern Botany.* New York: Charles Scribner's Sons, 1893.
[2] Letter to W. Ogle, 22 February 1882. In contrast, Peter Medawar, normally a reliable judge (p. 12) described Aristotle's works as a 'tiresome farrago of hearsay'.

understanding. Specimens and information sent to him from distant places conquered by his former pupil, Alexander (the Great), meant he could (and found it necessary to) look beyond his immediate neighbourhood in Greece. Centuries later, the cosy assumptions of medieval European scholars had to be rethought in the same way as specimens and descriptions from voyagers to distant parts began to pour into their world.

Aristotle was the first to recognize the value of comparing structures and processes in different organisms. He was also the first to study experimentally developmental processes and to describe the life histories of a large number of animals. He did not propose a formal biological classification, but his arrangement of invertebrates was superior in some ways to that of Linnaeus centuries later. He broke away from the suffocating embrace of his teacher Plato over the existence of a mysterious external and invariable force regulating nature.[3] He has left an enduring legacy by emphasizing the importance of asking 'Why?'.

Aristotle's emphasis on experiment and observation was not attractive to the early Christian theologians. Basil of Caesarea (329–79) argued that the behaviour of the elements must be understood in terms of laws ordained by God, rather than any thoughts about their essences; and that nature, once created and put in motion, evolves in accordance with its own laws, without interruption or diminishing of energy. Augustine of Hippo (354–430) completely and deleteriously divorced the natural order from the supernatural in his highly influential work *The City of God Against the Pagans*. He taught that the secular world was totally corrupt and worthless; for him the history of the world focused on a struggle between God and the Devil.

A few of Aristotle's writings were translated into Arabic and influenced great Islamic scholars like Avicenna (980–1037) and

---

[3] Ernst Mayr (*Growth of Biological Thought*. Cambridge, MA: Harvard University Press, 1982, p. 305) describes Plato as having 'a particularly deleterious impact on biology for two thousand years' through four dogmas – a belief in a constant *eidos*; an animate, harmonious cosmos; a creative demiurge; and a stress on a 'soul'.

Averroes (1126–98), but his reputation only resurfaced significantly in the twelfth to fourteenth centuries AD when his works began to be available in Latin and read more widely. Even then they were initially dismissed by the Church because they seemed to discount the notion of a personal God and an immortal soul; some religious authorities went so far as to ban his books. The god of whom Aristotle wrote seemed to be an impersonal agent concerned with teleology – a god who was a final agent, not an efficient one. His rehabilitation began when Albertus Magnus (1200–80) digested and interpreted Aristotle's writings in a way sympathetic to Christian thinking at the time, partly by treating the natural world as outside the concern of theology. He kept his religion and science in separate compartments. His pupil, Thomas Aquinas (1225–74), felt that reason and faith must be consistent with each other, but he also could not free himself from the conviction that they occupied different spheres.

Aquinas's *Summa Theologica* was – and is – highly influential, albeit more to theologians than other scholars. But two centuries later five naturalists born within a few years of each other, and contemporaries of Copernicus and Vesalius, did more than anything to reinvent an interest in the natural world and produce an upsurge in observational biology in the Middle Ages: William Turner (1508–68), father of English botany; the French traveller and explorer, Pierre Belon (1517–64); Guillaume Rondelet of Montpelier (1507–66); Ulisse Aldrovandi (1515–65), creator of the Botanic Garden in Bologna; and, most importantly, Konrad Gesner (1515–65), whose 4000-page *Historia Animalium* was a truly comprehensive collation of knowledge at the time. It was the first attempt by anyone to separate myth and tradition from observed facts. He described accurately many animals – although he also included mermaids and unicorns.

Gesner was a Renaissance man. Alongside his scientific interests, he had strong theological connections. He was born in Zurich and knew the Swiss Reformers personally during the height of the Reformation years. His father was killed fighting alongside

Zwingli in the conflict between the Protestant and Catholic cantons of the Swiss confederacy when young Konrad was fifteen. At one time he seemed destined to be a pastor, but he studied for an MD in Basel and then settled down to medicine and science in Zurich. He was a man with a wide perspective. He wrote, 'I have determined as long as God gives me life, to ascend one or more mountains every year when the plants are at their best – partly to study them, partly for exercise of body and joy of mind... I say then that he is no lover of nature who does not esteem high mountains very worthy of profound contemplation.'[4]

Gesner was best known among his contemporaries as a botanist. The century after his time saw great advances in plant biology, but progress in zoology was held up until structure replaced function or habitat as the criterion for classification. An exception to this zoological stasis was William Harvey (1578–1657), who dissected many animals and followed Vesalius in demonstrating the errors in Galen's generally accepted teaching; Harvey showed that the heart is essentially a fluid-circulating pump. He served as a link to the future by strengthening the idea that mechanism was a key element in understanding the natural world. This was also the time when Galileo was making himself unpopular with the Vatican because of his attempts to point out that scripture had to be reinterpreted because of astronomical discoveries. It has been said that it was Galileo's telescope, not his Bible, that conclusively refuted the assumption that 'the world is fixed immovably' (Psalm 96:10) – that the biblical words meant something theological rather than cosmological. Or, as Galileo's said, 'The Bible teaches us how to go to heaven, not how the heavens go.' Unfortunately, authority still had a higher priority than observation to Galileo's critics. It was a dogma that had to go. The pre-eminence of dogma was explicitly challenged by John Ray (1627–1705) and Francis Willughby (1635–73) in their *Ornithologiae libri tres* (1676), published after Willughby's early death.

---

[4] Letter to Jacob Vogel on the admiration of mountains. In: *Libellus de Lacte* (1545).

Ray was one of three remarkable Cambridge scholars who published books within ten years of each other at the end of the sixteenth century, which set the scientific agenda for the next hundred years. The best known was Isaac Newton (1643–1727), who produced his *Philosophiæ naturalis principia mathematica* in 1687 in response to importuning by the astronomer Edmond Halley (1656–1742), who wanted to know how Newton had solved the problem of planetary motion. Newton's work postulated universal gravitation and laid out a new science of dynamics, linking Kepler's laws of planetary orbits with Galileo's kinematics of terrestrial motion. The second of the three was Thomas Burnet (1635–1715), Fellow of Christ's College, Cambridge, and then Master of the Charterhouse in London. He believed that the original creation had been disastrously damaged by a global flood, with – his calculation – eight times the volume of water in the contemporary oceans bursting from hitherto unsuspected subterranean caverns and carving out the mountains and hills of our present world from the perfect sphere he believed God had created. He set out these ideas in *The Sacred Theory of the Earth* (1684; originally published in Latin as *Telluris theoria sacra* [1681]).

Burnet began with the almost unquestioned assumption of his time that God had created the world as a perfect sphere with a series of concentric layers surrounding it, the surface of the Earth lying as a crust lying over water, and 'no Rocks or Mountains, no hollow caves, nor gaping Channels, but even and uniform all Over'. Rivers ran from the poles to the tropics, where they dissipated. This primitive order disappeared in the devastation produced by the biblical flood. For Burnet, the Earth of his time was the shattered ruin of a 'perfect' creation before the flood; to him the oceans were gaping holes and the mountains upturned fragments of the crust of the original Garden of Eden. He was particularly offended by mountains: 'If you look upon a heap of them together they are the greatest examples of confusion that we know in Nature; no tempest or earthquake puts things into more disorder.' Burnet's book greatly impressed Newton, who wrote to Burnet to congratulate him. It was reprinted repeatedly throughout the 1700s and was regarded as a significant geological text. It was the most popular geology of the eighteenth century. Joseph Addison took his idea of the sublime from Burnet (p. 123).

Newton, Burnet and Ray were all concerned with God, but they reflected markedly different ways of picturing the interface between divine activity and the natural world. Leibniz accused Newton of supposing God was a bungling craftsman who could not make a perfect clockwork mechanism that would run without his attention. Notwithstanding, they all reflected an ushering-in of a modern phase of this meeting point. Speculations about how God(s) related to creation must be as old as conscious thought, but every such speculation has had to be constantly revised and refined as knowledge of real phenomena grows. In the sixteenth century the power of astrology and magic began to lessen, but there was a persisting habit of proposing a God that must be acting in places where there was a gap in knowledge.[5] Copernicus had destroyed Ptolemy's cosy cosmos and Vesalius had set in train a discipline of observation and experiment. Francis Bacon (1561–1626), William Harvey (1578–1657), René Descartes (1596–1650) and others provided reasons for regarding the world as a 'Great Machine', driven by laws and physical causation. Everywhere there appeared to be order and design, rather than chaos. In the words of John Locke (1632–1704), 'the works of Nature, in every part of them, sufficiently evidence a Deity'.[6] Metaphysical and occult explanations were no longer sufficient – or necessary. Notwithstanding, a 'Great Machine' does not do away with the need for a 'Divine Designer'.

---

[5] Henry Drummond (*The Ascent of Man*. Glasgow: Hodder & Stoughton, 1904, chapter 10) chastised those who pointed to things that science could not yet explain – 'gaps which they will fill up with God' – and urged them to embrace all nature as God's – the work of '... an immanent God, which is the God of Evolution, is infinitely grander than the occasional wonder-worker, who is the God of an old theology'. The inadequacy of such a God-of-the-gaps was powerfully exposed by Oxford chemist Charles Coulson in *Science and Christian Belief* (Oxford University Press, 1955). It still remains regrettably fashionable in some circles as 'intelligent design'.

[6] *The Reasonableness of Christianity, as Delivered in the Scriptures*, 1695.

The three Cambridge men all had their personal theological agendas as well as scientific ones, and this brew influenced both natural philosophy and what came to be called 'physico-theology' for at least the next century (p. 76). John Ray's *Wisdom of God Manifested in the Works of Creation* appeared four years after Newton's *Principia*. Ray was more orthodox than either Burnet or Newton. In his passion to interpret the Bible rigorously, Burnet had to speculate well beyond contemporary knowledge, while Newton was increasingly obsessed with alchemy and interpreting biblical prophecies. Newton believed in a God who intervened in the natural world, but he came to deny the divinity of Jesus. His posthumously published *Chronology of Ancient Kingdoms Amended* (1728) and *Observations on Daniel and the Apocalypse of St. John* (1733) were reconstructions of biblical chronology in the light of astronomical data, which Newton believed had been falsified by popery and priestly machinations.

In contrast, Ray was essentially a naturalist, like Darwin two centuries after him. His biological writings were firmly based on observation, and he extended the same discipline into his theology. His *Wisdom* makes two important and lasting generalizations. Ray captured the growing acceptance that God had wider concerns than humankind. He wrote, 'It is a generally received opinion that all this visible world was created for Man [and] that Man is the end of creation, as if there were no end of any creature but some way or other to be serviceable to man... But though this be vulgarly received, yet wise men nowadays think otherwise.' This was a bold statement, and it certainly did not mark the end of anthropocentrism. Indeed, seventeenth-century scholars treated the history of the Earth as if it was human history, albeit told to us by divine 'revelation'. For them, the non-human world was little more than the setting for the human story. Only occasionally did God intervene in the natural world – such as when the Red Sea parted to enable the Israelites to make their exodus from Egypt and gain their freedom. However, there were two points where the natural world did figure in the forefront of the sacred story: creation itself and the universal flood of Noah.

BOX 4.1 **The biblical flood**

Noah's flood provides a topic of endless speculation. The Bible story of Noah tells of an ark built at God's command, into which Noah took in 'a male and a female of all beasts, clean and unclean, of birds, and of everything that creeps on the ground' to save them from drowning in an imminent deluge (Figure 4.1). The interpretation of this story has exercised biblical exegetes over many centuries: Could every species fit into the ark? Where was the fodder stored? What happened to the dung? How could carnivores and herbivores coexist? What about the parasites? The questions grew worse in the seventeenth century as the New World became increasingly known. How should the Americas be understood: as a land that had wholly escaped from the flood? A sodden continent only recently risen above

FIGURE 4.1 Noah's ark on Mount Ararat. Painting by the Flemish artist Simon de Myle, c.1570, courtesy of Heritage Images. The interpretation of the flood story became ever more complicated in medieval times.

the waters? Or perhaps the New Atlantis described by Plato in his dialogue *Timaeus*?

Even lawyers tried their hand at explanations. Lord Chief Justice Matthew Hale (1609–76) proposed that animal life in America might have been the result of a kind of migration and subsequent degeneration – or, if not, maybe American species are really worldwide and they would be found elsewhere when Africa and Asia were more thoroughly explored. He suggested that Native Americans might be descendants of the lost tribes of Israel, or perhaps naked innocents who had escaped the fate of Noah's wicked generation, or even degenerate savages who had wandered away after the ark grounded. In France, Isaac de La Peyrère (1594–1676) argued that the flood was a local event and that Adam was not the first-created man, but merely the first Jew, while the natives of the Americas were another species altogether. Hale rejected this as 'immoral and irreligious' and insisted that all humans were descended from Adam, and that the same argument should apply to the animal kingdom: all animals were descended from those that came out of the ark.

A century after Hale and La Peyrère, Linnaeus listed fourteen thousand species, five and a half thousand of them animals. How could so many creatures have been passengers on the ark? Linnaeus was not afraid of sticking his neck out. In 1748 he was summoned before the Faculty of Divinity at Uppsala University where he taught, to defend a treatise he had written on the relationship between natural science and religion. Although he seems to have treated the occasion respectfully, he apparently took umbrage at the attitude of the theologians and responded with a fourteen-point manifesto. He regarded himself as 'the publisher and interpreter of the wisdom of God' and opted for a revisionist version of the Bible. He suggested that all living beings, including humankind, originated on a high mountain at the time the primeval waters were beginning to recede. Extrapolating backwards, as it were, he proposed that in the beginning only one small island had been raised above the surface of a worldwide sea. This must have been the site of Paradise and the first human home.

As the mountain grew higher and higher, he argued that it would have presented a wide range of ecological conditions, arranged as belts from tropical to polar zones. He envisaged organisms moving to find the latitudes where they were to remain for the rest of time. He did not contest the biblical account of creation, but questioned the ark episode; it did not fit into his understanding. He asked, 'Is it credible that the Deity should have replenished the whole earth with animals to destroy them all in a little time by a flood, except a pair of each species preserved in the Ark?'[1] He believed it was a combination of sheer chance, the speed of migration and the helping hand of God that determined exactly where each species ended up.

[1] Linnaeus, C. *Dissertation on the Increase of the Habitable Earth* . London: G. Robinson and J. Robson, 1744.

Ray's other generalization was expressed in the preface of his *Ornithology* (1678). Writing with his friend and patron Francis Willughby, he declared they had 'wholly omitted what we find in other Authors concerning *Homonymous* and *Synonymous* words, or the divers names of *Birds, Hieroglyphics, Emblems, Morals, Fables, Presages* or ought else pertaining to *Divinity, Ethics, Grammar*, or any sort of Humane Learning'. In other words, Ray was concerned to distance himself from the convention that living things were little more than symbols with an underlying meaning, choosing to concentrate instead on 'only what properly relates to their Natural History' – an attitude which we would nowadays unhesitatingly accept as appropriate (Figure 4.2).

Someone else who illustrates well the spirit of this period is the physician and polymath Sir Thomas Browne (1605–82). His aim was to see life as a whole, to find an intimate correspondence between the world of experience and inherited traditions – and to interpret them coherently, so that thought and conduct might be integrated into a consistent unity of achievement. He has been described as

FIGURE 4.2 John Ray (1627–1705), 'the father of natural history', who distinguished observation from allegory in the study of nature, and thus opened the way forward for the objective recording of Gilbert White and many others.
Print by A. de Blois after painting by William Fairborne.
Photo: Science & Society Picture Library/Getty Images.

representing 'the coming of modern man',[7] although a more accurate description might be that he was one of the last Renaissance men – able to bring together the whole range of human knowledge in a way that was becoming increasingly difficult. Browne consciously linked the piety of earlier times with the intellectual discipline of the

[7] Raven, C.E. (1947). *English Naturalists from Neckam to Ray.* Cambridge: Cambridge University Press, p. 342ff.

seventeenth century – insisting on the right and duty of enquiry, the competence and responsibility of human reason and the application to all phenomena of the method of verification and induction. In his 1646 book *Pseuodoxica Epidemica* (commonly known as 'Vulgar Errors'), he took it upon himself to expose common misconceptions, such as that 'an elephant has no joints', 'a horse has no gall', 'a beaver bites off its testicles to escape a hunter'. He believed that we live in two 'divided and distinguished worlds' (he coined the phrase 'Man the great Amphibium' to express this idea) and refused to accept any contradiction between faith and reason. His best-known work is *Religio Medici* (1643). He wrote, 'There are two bookes from whence I collect my Divinity: besides that written one of God, another of his servant Nature, that universall and publik Manuscript, that lies expans'd unto the eyes of all; those that never saw him in the one, have discovered him in the other. This was the Scripture and Theology of the Heavens.' Browne lived at the junctions of two epochs: a double-faced age, half scientific and half magical; half sceptical, half credulous. He believed in the Devil in a way reminiscent of John Milton. He regarded the mechanical view of nature as insufficient for it to function effectively, but he wanted to reject unsupported claims of infallibility and to expose superstition. He combined (albeit incompletely) the world of observed fact with that of inferred meaning, arenas which had been separated by the medieval failure to relate symbolism to reality.

Browne's successors narrowed this understanding of knowledge. The eighteenth century was the 'Age of Reason'. In his dissection of it, Roy Porter identified the key concept as 'Nature' – more specifically, the deification of Nature:

> Deeply enigmatic, it is most easily approached in terms of its opposites. It was an affirmation of an objective and external reality, created by God, repudiating the fallen, decaying cosmos imagined by Calvinism. The natural world could also serve as the antithesis of all that was confused and contorted, the deceitful and the meretricious. For early environment thinkers like [the 3rd Earl of]

Shaftesbury, Nature linked the divine (eternal and transcendental) and the human; it pointed to the purification and perfection of mankind, and extended human sympathies beyond the normal bounds of artifice. Orderly, objective, rational, grand and majestic Nature enshrined both norms and ideals.[8]

One word which may be taken as summing up this attitude was 'physico-theology' often taken as equating to 'natural theology' – the idea that evidence for God's existence can be found in the natural world by the use of unaided (natural) reasoning, complementing 'revealed theology' (or 'revealed religion'), which is based on God's revelation of himself in scripture and religious experiences. The term 'physico-theology' seems to have been first used in 1652 by Walter Charleton in the subtitle to his book *Darkness of Atheism Dispelled by the Light of Nature*, in which he argued that the ordering and beauty of the natural world clearly showed the existence and nature of God. Ray himself used the word in the title of the second edition of his *Miscellaneous Discourses Concerning the Dissolution and Changes of the World*, published in 1692, the year after *Wisdom*. A second edition appeared only a year after the first, but with a new title: *Three Physico-theological Discourses*. This book was an expansion of a sermon preached thirty years earlier at 'S. Maries church in Cambridge' (presumably Great St Mary's) on the biblical text: 'The day of the Lord will come like a thief. On that day the heavens will be dissolved in flames, and the earth and all that is in it will be brought to judgement. Since the whole universe is to dissolve in this way, think what sort of people you ought to be, what devout and dedicated lives you should live' (2 Peter 3:10–11).

This harked back to Burnet's *Sacred Theory*. However, the main interest in Ray's *Miscellaneous Discourses* was his insistence that fossils were the remains of living animals. He retracted this belief in the final edition of the book in 1713, apparently because he was theologically troubled by the implications of extinction. How could God allow this?

[8] *Enlightenment*. London: Allen Lane, 2000.

BOX 4.2 **The fossil record**

Fossils were well known in antiquity. They were believed to be either natural artefacts (or 'crystals') or the remains of creatures which had perished in Noah's flood. Many of the latter 'more or less' resembled living animals or plants, but this identity was increasingly challenged, notably by Nicolas Steno (1638–86), an anatomist who became a bishop and who contended on the basis of the rocks and fossils he studied in Tuscany that it was possible to reconstruct a historical sequence of natural events, a 'fossil record'. Another challenger was Steno's contemporary, Robert Hooke (1635–1703), Curator of Experiments for the Royal Society in London, who studied fossils with a microscope and concluded that they were indeed the remains of living things that could give clues to the past history of life on Earth (Figure 4.3).

By the end of the eighteenth century those familiar with the evidence had no doubt that the Earth had a much longer history than the few millennia previously assumed. Many scholars had attempted to describe this history, the best known being Archbishop Ussher of Armagh, who in 1650 made a calculation of 4004 BC for the date of creation, a date which for many years was inserted at the beginning of English bibles. Various geological epochs were recognized, although their absolute lengths were unknown until the application of radiometric dating in the early decades of the twentieth century. But there was no doubt that there was a procession of change and that the changes shown in the rocks and their fossils must have involved many thousands of years. Firm evidence of these changes was collected by an English surveyor, William Smith (1769–1839), and independently by a French zoologist, Baron Georges Cuvier (1769–1832). They showed that particular rock strata have distinctive fossils. Smith was involved in canal-building and tracing coal seams for mining, and he realized that geological strata could be characterized by the fossils in them. Such strata can often be followed for hundreds of miles, even when the rock formation changes. Smith developed these data between 1791 and 1799 and produced a 'stratigraphic map' of England and Wales, although this was not published until 1815. During the same period, French naturalists were collecting fossils in the limestone

Geological Column

		Millions of years (mya)		
CENEZOIC ERA (Age of recent life)	Quaternary period			Ice ages and modern times
		1.8	The Tertiary period can be divided into 5 epochs - a younger Tertiary with Miocene and Pliocene epochs, during which the first hominids appeared; and an older Tertiary with Palaeocene, Eocene and Oligocene epochs.	
	Tertiary period			Age of mammals
		6.5	Derived from Latin word for chalk (creta) and first applied to extensive deposits that form white cliffs along the South coast of England	
MESOZOIC ERA (Age of dinosaurs)	Cretaceous period			Last dinosaurs, first primates, first flowering plants
		14.1	Named for the Jura mountains, located between France and Switzerland, where rocks of this age were first studied.	
	Jurassic period			Dinosaurs dominant, first birds
		19.5	Taken from the word 'trias' in recognition of the threefold character of these rocks in Europe.	
	Triassic period			Appearance of dinosaurs
		23.0	Named after the Russian province of Perm, where these rocks were first studied.	
	Permian period			Extinction of many marine forms
		28.0	Named after Devonshire, England, where these rocks were first studied.	
	Carboniferous period			First reptiles, seed ferns
		34.5	Named after Celtic tribes, the Silures and the Ordovices, that lived in Wales during the Roman Conquest.	
PALEOZOIC ERA (Age of ancient life)	Devonian period			First amphibians, jawed fish
		39.5	Taken from the Roman name for Wales (Cambria) where rocks containing the earliest evidence of complex forms of life were first studied.	
	Silurian period			First vascular land plants
		43.5		
	Ordovician period			
		50.0		First fish
	Cambrian period			
		57.0	The time between the birth of the planet and the appearance of complex forms of life. More that 80 per cent of the Earth's estimated 4-1/2 billion years falls within this era.	
PRECAMBRIAN	Proterozoic archean			

FIGURE 4.3 Geological column and the fossil record

quarries around Paris. From these, Cuvier worked out the detailed stratigraphy of these fossils (mainly mammals). The conclusion, unpalatable as it was at the time, was that there is a time sequence involved in the laying down of fossil-bearing strata, with the lowest strata being the oldest. Later it became possible to correlate strata, not only across England or western Europe but between different parts of the world.

The ignorance of dating did not affect the acceptance by the early 1800s of the fossil record as an accurate, albeit incomplete, history of life.

The fascination in the eighteenth century with physico-theology was driven by Newton's mechanics rather than Ray's biology. Newton put forward a vision of the natural world which reinforced the rationalism of doctrine of the Latitudinarian theologians in the ascendancy at the time. Their arguments were fashionable, not least because they supported the reconciliation of divine and social ordering as represented by the Glorious Revolution of 1688 with the ejection of King James II and the repudiation of his belief in a divine right to rule. Physico-theology found a natural expression in the Boyle Lectures, established by the will of the pioneering chemist, Robert Boyle (1627–91), to provide a salary for a 'learned Minister of the Gospel residing in the City of London... to preach Eight Sermons in the Year for proving the Christian Religion against notorious Infidels, viz. Atheists, Theists, Pagans, Jews, and Mahometans'. The first series was delivered in 1692 by Richard Bentley, a classicist who later became Master of Trinity College. Newton approved. He wrote to Bentley, 'when I wrote my treatise about our Systeme I had an eye on such Principles as might work with considering men for the beliefe of a Deity & nothing cn rejoyce me more than to find it usefull for that purpose'. Following a conversation with Newton in 1691, the Scottish astronomer David Gregory (1659–1708) recorded Newton's confidence that 'the most simple laws of nature are observed in the structure of a great part of the Universe, that the philosophy ought there to begin, and that Cosmical Qualities are as much more Universall than particular ones, and the general contrivances simpler than that of Animals plants etc'. Put another way, Newton did not seem too impressed with Ray's recently published *Wisdom*.

Any neglect of Ray in the early Boyle Lectures was more than redressed by William Derham in his 1711 and 1712 lectures, published in 1713 as *Physico-theology or a Demonstration of the Being and Attributes of God from His Works of Creation*. Derham was the Vicar of Upminster in Essex and a Fellow of the Royal Society. It was said that his book contained 'little intellectual excitement'. The book was dedicated to Ray and used much material from *Wisdom*, as well as observations made by Derham himself.

Other eighteenth-century biologists were explicit in their gratitude to Ray. Gilbert White (1720–93), author of the *Natural History of Selborne* (1789), regarded Ray as his mentor, both scientifically and theologically (Figure 4.4). He extolled Ray in a letter to Daines Barrington of 1771 as 'Our countryman, the excellent Mr Ray [who] is the only describer that conveys some precise idea in every term or word, maintaining his superiority over his followers and imitators, despite of the advantage of fresh discoveries.' Even John Wesley, often accused of exciting enthusiasm at the expense of rational thought, drew up and revised over fifteen years or so a *Survey of the Wisdom of God in the Creation* to interpret science to 'common readers', a survey which included notes from Ray and Derham. He expressed his attitude in words which Ray himself might have used: '[God] does not impart to us the knowledge of himself immediately; that is not the plan he has chosen; but he has commanded the heavens and the earth to proclaim his existence, to make him known to us. He has endued us with faculties susceptible of this divine language, and has raised up men who explore their beauties and become their interpreters.'

We now come to the most well-known biologist of the eighteenth century, the Swede, Carl Linnaeus (1707–78). He had a much narrower vision than Ray. Classification for Ray was merely

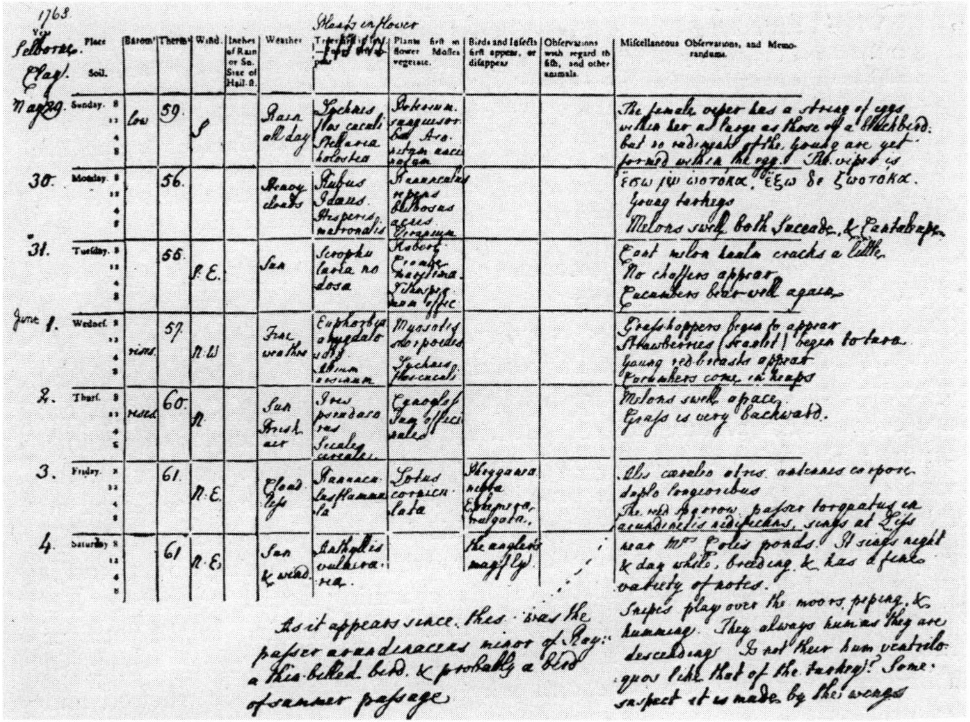

FIGURE 4.4 A table from Gilbert White's journal listing weather with plants in flower. White acted as the 'curate' of Selborne in Hampshire on four occasions (the first for his uncle, who was the vicar) and kept detailed records of his observations in his garden and on his travels and in the area. These were eventually edited together as a series of letters into *The Natural History of Selborne* (1788), the fourth most published book in English.

Photo: Culture Club/Getty Images.

a tool for clarification of biological situations; for Linnaeus, the fundamentals of biology were simply classification and nomenclature. Notwithstanding, Linnaeus set a basis for biological realism by his collections, derived from systematically soliciting specimens from a wide range of localities. He spread so-called 'apostles' in many parts of the globe. Probably the most famous of these was Daniel Solander (1733–82), whom Linnaeus was grooming to be his successor. However, Solander, sent to England to publicize Linnaeus's ideas, stayed there. Eight years after his arrival (in 1768), he was employed to accompany Joseph Banks (1743–1820) on the first voyage of the *Endeavour* under Captain James Cook, one of the most important voyages of discovery ever made.

Banks bestrode British science for half a century. He was President of the Royal Society for forty-two years, friendly with royalty, unofficial director and re-invigorator of the Royal Botanic Gardens at Kew, patron of many scientists and scientific expeditions, and energetic advocate of transplanting economically important species. Specimens collected by Banks led to the recognition of 110 new genera and around 1300 new species. Despite all this, his main legacy is as an administrator rather than a practising biologist. He has been properly called 'a statesman of science'.

A younger, German contemporary of Banks, Alexander von Humboldt (1769–1859), was inspired by Banks and had a much greater influence on science *sensu stricto*. He travelled extensively in Central America and northern South America, fired by his Göttingen University friend, Georg Forster (1754–94), who had gone with his father as a naturalist on James Cook's second voyage (replacing Banks, whose demand for accommodation for his eighteen-strong entourage was so excessive that Cook refused him a place on the ship).[9] Besides

---

[9] Cook does not appear to have been enamoured of his naturalists. He refused to take one on his third voyage, declaring, 'Curse scientists – and all science into the bargain.' Banks took umbrage at being excluded and went to Iceland instead, recommending to the British Government that they should annex Iceland for the Empire.

the accounts of his travels, Humboldt laid the foundations of both physical and botanical geography. On his way across the Atlantic, he landed on the Canary Island of Tenerife and described vertical zonation in the vegetation of the central mountains, the sort of analysis which nowadays forms the preliminary stage of any ecological study. He extended this work in South America, most famously on Chimborazo, which was then thought to be the highest mountain in the world. He invented isobars and isotherms as an aid to showing the limits of particular species and of natural assemblages. Perhaps even more importantly, he developed 'botanical arithmetic' and thus laid the basis for quantitative analysis in ecology. Humboldt's botanical arithmetic was simply the ratio of species in one group of plants to that in another. This can show the predominant forms present in a region and the general relationships between different groups. The technique was developed further by Augustin de Candolle (1778–1841) and his son Alphonse (1806–93); it was used by Leopold von Buch (1774–1853) in his essay on the Canary Islands where he suggested isolation was a prerequisite for speciation (p. 96), by Charles Darwin in comparing the floras of different archipelagos and by Darwin's friend Joseph Hooker (1817–1911) in comparing 'continental' and 'insular' populations.

Humboldt inspired a generation of naturalists. He was a hero to Darwin, who read Humboldt's *Personal Narrative* while at Cambridge and took with him on his five-year voyage on the *Beagle* an edition of Humboldt's travels inscribed, 'J.S. Henslow [Darwin's friend and botanical teacher in Cambridge] to his friend C. Darwin on his departure from England on a voyage round the World, 21 Sept 1831'. He repeatedly refers to Humboldt in his *Beagle* journal. Acutely seasick while crossing the Bay of Biscay early in the voyage, he wrote, 'I am at present fit only to read Humboldt; he like a sun illumines everything I behold... Nothing could be better adapted for cheering the heart of a sick man'. In his *Autobiography*, he described his first venture into the Brazilian forest, declaring, 'I formerly admired Humboldt, I now almost adore him; he alone gives any notion of the feelings which are raised in the mind on first entering the Tropics.'[10] Late in life, Darwin told Joseph Hooker that Humboldt was 'the parent of a grand progeny of scientific travellers'.

At the same time that Banks was sailing round the world with James Cook and Humboldt was preparing to cross the Atlantic, a rather mild clergyman, William Paley (1743–1845), was a Fellow and teacher of moral philosophy at Christ's College Cambridge (1766–76). He then moved to incumbencies in the Diocese of Carlisle, becoming Archdeacon there in 1782. His bishop persuaded him to publish his Cambridge lectures. These appeared in 1785 under the title *The Principles of Moral and Political Philosophy*, and were an instant success. A later version, *A View of the Evidences of Christianity* (1794), was required reading for Cambridge undergraduates until 1920. A more popular book, *Natural Theology or Evidences of the Existence and Attributes of the Deity*, followed in 1802. Paley had been deeply impressed by Newton's demonstration of the regularity of nature. The image of the world as a mechanism suggested to him the metaphor of a clock. For Paley, the world was full of beautifully functioning biological structures which must have been 'contrived' – in other words, someone must have made them. This implied to him a purpose in their existence and hence something about the contriver who had designed and made them. He transposed the fashionable arguments of the eighteenth century for the existence of God from physical to biological design. He was certainly not an original thinker,[11] but he was very influential during an age when machinery was becoming increasingly important in national life.

Paley's arguments did not go unchallenged. For many, *Evidences* was not even really Christian. Paley based his notions of right and wrong on purely natural reasoning, 'confident that all men could be brought to agree' – a methodology commonly

---

[10] *The life and letters of Charles Darwin, including an autobiographical chapter.* London: John Murray, 1887.

[11] Most of the examples used by Paley were lifted (without acknowledgement) from Ray's *Wisdom of God Manifested in the Works of Creation*.

adopted by tabloid journalists. His thinking was anathema to the Romantics, who were becoming increasingly influential. Samuel Taylor Coleridge complained about '... the prevailing taste for books of Natural Theology, Physico-Theology, Demonstrations of God from Nature, Evidences of Christianity and the like... *Evidences* of Christianity! I am weary of the word! Make a man feel the *want* of it; rouse him, if you can, to the self-knowledge of his *need* of it; and you may safely trust it to its own evidence.'[12] There were efforts as early as 1830 to remove Paley's books from the compulsory reading list at Cambridge.

It is customary to identify Paley as the key proponent of a biological stasis which was overturned by Darwin. Paley's belief was the unshakeable principle – to him – of a creator God who was a Great Artificer or Watchmaker, responsible for all the 'perfect adaptations' and 'contrivances' which Paley found everywhere in the natural world. He saw all these features as specially created by God, fitting every living being into an overall divine plan. Darwin began by being heavily influenced by Paley; he re-read *Natural Theology* in 1843 when at the height of developing his evolutionary ideas – a time between first writing them down in a 'Sketch' in 1842 and expanding them into an 'Essay' in 1844. He knew Paley's writings well, commenting in his *Autobiography*, 'The careful study of these works, without attempting to learn any part by rote, was the only part of the Academical Course [in Cambridge] which I then felt and still believe, was of the least use to me in the education of my mind.' He wrote that, in his degree course, Paley's logic 'gave me as much delight as did Euclid', although he qualified this memory: 'I did not at the time trouble myself about Paley's premises; and taking these on trust, I was charmed and convinced by the long line of argumentation.'

But Paley's comfortable teleology, describing a harmonious world created and maintained by a benevolent God was a world-view under stress even as Paley wrote. His contemporary, William Herschel

---

[12] *Aids to Reflection.* London: William Pickering, 1848.

(1738–1822), was showing the inadequacy of Newton's static universe with his astronomical observations. The early geologists were describing changes in the Earth's rocks which must have occurred over very much longer periods than the 6000 to 10,000 years assumed from biblical chronologies. This growing realization of a long Earth history, supported by the observations of James Hutton (1726–97), Charles Lyell (1797–1875) and others, allowed enough time for the procession of evolutionary change assumed by the French biologist, Jean-Baptiste Lamarck (1744–1829), but sounded a death knell for traditional natural theology. A creator could presumably design an organism perfectly adapted to a particular environment, but this perfection would disappear if the environment changed. Adaptation to any changes – in climate, to the physical structure of the Earth's surface or to predators and competitors – is possible only if organisms themselves change. The passage of time has been described as 'the Achilles heel of natural theology'. On top of this there were growing problems from attributing all animal distribution to spread from Noah's ark.

A last throw for Paleyian-ism was the bequest, allegedly in attempted expiation for a misspent life, of the Reverend Francis Egerton, eighth Earl of Bridgewater, to pay eight scientists £1000 each to examine 'the Power, Wisdom, and Goodness of God, as manifested in the Creation'. The resulting eight books, published between 1833 and 1837, repeated with prolix but best-selling tedium the arguments of John Ray as mangled by Paley. But knowledge of the world had moved on. The static world of earlier times was disappearing under a torrent of technological innovation and scientific discovery. While the eight selected authors – Chalmers, Kidd, Whewell, Bell, Roget, Buckland, Kirby and Prout – were writing their Bridgewater Treatises, Charles Darwin was sailing round the world on HMS *Beagle* (1831–6). He opened his first notebook on 'transmutation' in the early summer of 1837. He was sufficiently sure of himself by 1842 to write a 35-page 'Sketch', which he expanded into a 200-page 'Essay' in 1844; this 'Essay' formed the basis of *On the Origin of Species* fifteen years later.

Cambridge theologian Don Cupitt has well described the intellectual dilemma of the time, reflecting the angst of Coleridge and others:

> Mechanistic science was allowed to explain the structure and workings of physical nature without restriction. But who designed this beautiful world-machine and set it going in the first place? Only Scripture could answer that question. So science dealt with the everyday tick-tock of the cosmic framework and religion dealt with the ultimates: first beginnings and last ends, God and the soul... It was a happy compromise while it lasted. Science promoted the cause of religion by showing the beautiful workmanship of the world... But there was a fatal flaw in the synthesis. Religious ideas were being used to plug the gaps in scientific theory. Science could not yet explain how animals and plants had originated and had become so wonderfully adapted to their environment – so that was handed over to religion. People still made a sharp soul–body distinction and the soul fell beyond the scope of science – so everything to do with human inwardness and persona and social behaviour remained the province of the preacher and moralist.[13]

BOX 4.3   **Charles Darwin**

Darwin (1809–82) has to be regarded as a key agent in influencing environmental attitudes. He was an unlikely candidate. At school, he was considered 'a very ordinary boy, rather below the common standard in intellect'. His father berated him, 'You care for nothing but shooting, dogs and rat-catching.' He dropped out of medical school in Edinburgh, moving on to do a general degree in Cambridge. After his time at the Universities of Edinburgh (1825–7) and Cambridge (1828–31), he spent five much more formative and exciting years (1831–6) as a 'gentleman naturalist' on HMS *Beagle*,

---

[13] Cupitt, D. (1984). *The Sea of Faith*. London: BBC, p. 59.

commissioned under the command of Robert Fitzroy to survey the southern coasts of South America. Darwin began by treating this as a sort of extended gap year, assuming that he would seek ordination to the Anglican ministry on his return. He wrote home at an early stage of the *Beagle*'s voyage, 'Although I like this knocking about, I find I steadily have a distant prospect of a very quiet parsonage & I can see it even through a grove of Palms.' But during his time on the *Beagle*, he began to drift away from the idea of career as a clergyman, although he never renounced his faith.

When Darwin got back from his five years on the *Beagle* he stayed in London, busying himself sorting and publishing the data he had collected on the voyage. Three years later (in 1839) he married his cousin Emma Wedgwood. They lived briefly in London, but moved in 1842 to Down House, a square brick building built towards the end of the previous century, close to the small village of Downe, near Bromley in Kent. There the Darwins lived for the rest of Charles's life. He never again left Britain and travelled only occasionally. He died in 1882, expecting to be buried in the local parish church, but was instead interred in Westminster Abbey through the manipulations of his friends in the X Club (p. 140).

Darwin wrote of himself in his *Autobiography*:

> I have no great quickness of apprehension or wit which is so remarkable in some clever men, for instance [Thomas Henry] Huxley. I am therefore a poor critic: a paper or book, when first read, generally excites my admiration, and it is only after considerable reflection that I perceive the weak points. My power to follow a long and purely abstract train of thought is very limited; I should never have succeeded with metaphysics or mathematics. My memory is extensive, yet hazy: it suffices to make me cautious by vaguely telling me that I have observed or read something opposed to the conclusion which I am drawing, or on the other hand in favour of it; and after a time I can generally recollect where to search for my authority... On the favourable side of the balance, I think I am superior to the common run of men in noticing things which easily escape attention, and in observing them carefully. My industry has been nearly as great as it could have been in the observation and collection of

> facts. What is far more important is that my love of natural science has
> been steady and ardent.
>
> Whether or not this was a fair analysis of his talents and limitations,
> the publication of the *On the Origin of Species* in 1859 was a true
> tipping point in the history of ideas.

Despite his original approval of Paley, Darwin could not evade the debates about geology and biogeography which were swirling around in the first half of the nineteenth century. It was a time that saw the traditional understanding that animals and plants were created in their contemporary form and spread from a single site increasingly strained. Charles Lyell tried to save the concept of a divine creation by proposing in the second volume of his *Principles of Geology* (1832) that there might have been multiple 'centres of creation', with different species being created (and becoming extinct) *in situ*. Part of his reason for this were the plants on oceanic islands. He suggested that 'the original isle was the primitive focus or centre of a certain type of vegetation, yet all belonging to the same group, giving the appearance of centres of foci of creation... as if there were favourable points where the creative energy has been in greater action than in others'.

Charles Lyell was one of Darwin's closest scientific friends in the post-*Beagle* years. Darwin took the first volume of Lyell's *Principles of Geology* (published in 1830) with him when he embarked on the *Beagle*; the second volume was published in 1832 and was sent out to the *Beagle*, reaching Darwin in Montevideo; the third volume (which appeared in 1833) got to him in the Falkland Islands in March 1834. Lyell's ideas loosened Darwin from the Platonic stasis of Paley and pushed him towards the recognition of geological change and long-operating mechanisms. Lyell became a colleague when Darwin returned to Britain where he was for a time Secretary of the Geological Society. Lyell went on to become a staunch friend and adviser, joining with the botanist Joseph Hooker to present Darwin's work to the Linnean Society in 1858. But Darwin was undoubtedly closer to Hooker (1817–1911); he described him as 'one of my best friends throughout life. He is a delightfully pleasant companion and most kind-hearted.' More than anyone else, he served as a sounding board and critic of Darwin from the time in 1843 when they first corresponded until Darwin's death in 1882.

Other influences on Darwin were his teachers at university, particularly Robert Grant (1793–1874) and Robert Jameson (1774–1854) in Edinburgh and Adam Sedgwick (1785–1873) and John Stevens Henslow (1796–1861) in Cambridge. Robert Jameson was Regius Professor of Natural History in Edinburgh for 50 years (1804–54). Before that, he had studied for two years (1802–4) under Humboldt's tutor, Abraham Werner (1750–1817) in Freiberg. This led to him becoming a major proponent in Britain of Werner's 'Neptunism', that the Earth's rocks are all sedimentary, deposited from water, a process hastened by Noah's flood. He was an energetic field-worker and described the geology of many of the Scottish islands. Jameson had the reputation of being an inspiring and enthusiastic teacher, although Darwin recorded that Jameson's lectures 'were incredibly dull'. Notwithstanding, Darwin must have been somehow intrigued by geology; he attended and enthused about Sedgwick's lectures on the subject when he got to Cambridge. Sedgwick reminded him of his heroes: Humboldt, the astronomer Herschel and Paley rolled into one. He accompanied Sedgwick on a field trip to North Wales in 1831, shortly before his years sailing around the world on the *Beagle*.

More important to him was Henslow, Professor of Botany at Cambridge from 1827 until his death in 1861, although from 1837 he was also Rector of the village of Hitcham, about 45 km east of Cambridge. He had a special place in Darwin's life and affections. At Henslow's home in Cambridge, Darwin met many of the local scientists (including Sedgwick); he became known as 'the man who walked with Henslow'. Henslow acted as the receiver for the specimens collected by Darwin and shipped back at intervals from the *Beagle*. Later in life, he became increasingly committed to his parish ministry and his contact with Darwin lessened. But he continued to teach at Cambridge and maintained his connections in the scientific world. The year before his death, he chaired the notorious confrontation between the Bishop of Oxford and Thomas Huxley at the 1860 meeting of the British Association for the Advancement of Science in Oxford. Darwin described his friendship with Henslow as the most important circumstance in his whole career; he wrote to Joseph Hooker (who married Henslow's daughter), 'I believe a better man never walked this earth.'

However, Joseph Hooker was Darwin's closest friend. Like Darwin he had travelled round the world as a naturalist. Indeed, his role model on his voyage was Darwin, although he also looked up to Humboldt. He regarded Humboldt as 'the founder' of biogeographical studies, the Manx zoologist Edward Forbes (1815–54) as their 'reformer' and Darwin as their 'latest and greatest lawgiver'. T.H. Huxley (1825–95), who acted as naturalist on board HMS *Rattlesnake* on a discovery and surveying voyage to Australia and New Zealand (1846–50), is sometimes linked to these pioneer travellers, but his contributions from his *Rattlesnake* years were much less, and largely in the realm of coelenterate taxonomy. He confessed in his *Autobiography* of 1889 that 'there is very little of the genuine naturalist in me. I never collected anything and species work was always a burden for me.'

Another influence on Darwin might be expected to be Jean-Baptiste Lamarck, a Frenchman who worked in the Natural History Museum in Paris from 1788 until his death in 1829, but his importance seems to have been small. Born a year after Paley, Lamarck's ideas were basically theological. He believed there had been a progressive increase in perfection from the simplest organisms to its pinnacle in humankind; he argued that, over a long period of time, one species would become transformed into another and 'higher' one. This got over the problem of species extinction, which seemed to contradict the notion of a perfect world created by God – an idea which was becoming increasingly contentious in the late eighteenth century as it became ever clearer that organisms found as fossils were not still roaming in some as yet undiscovered El Dorado.

Lamarck's evolutionary ideas were not widely accepted outside Germany and his native France, although in Britain they were welcomed by intellectual radicals as a justification for challenging the social status quo. Lamarck apparently contributed little to Darwin's thinking. In the 'Historical Sketch' in *On the Origin of Species*, Darwin refers to him as 'this justly-celebrated naturalist' and acknowledges his 'eminent service of arousing attention to the probability of all change in the organic, as well as in the inorganic world, being the result of law and not of miraculous interposition'. However, he had earlier written to Hooker (11 January 1844), 'heaven forfend me from Lamarck's nonsense of a "tendency to progression"' and later (10 December 1844), 'Lamarck's [book] is veritable rubbish'. He commented in a letter to Baden Powell (Professor of Mathematics at Oxford, one of the authors of *Essays and Reviews* and father of Robert Baden Powell, the founder of the Boy Scout movement), 'By the way, his erroneous views were curiously anticipated by my Grandfather' (18 January 1860). Lamarck was obviously not one of Darwin's heroes; his main importance is that he prepared the way for Darwin by pointing to evidence that evolutionary change must have occurred.

More positive influences on Darwin certainly included John Gould, the ornithologist at the London Zoo, to whom Darwin had entrusted the bird specimens he had collected while on the *Beagle*. Gould told Darwin that in his collection of birds from the Galapagos Archipelago there were many species present (not mere 'varieties', as Darwin had assumed); his apparently serendipitous reading of Malthus then showed him the significance of the widespread 'struggle for existence' in the natural world.

Then there was the Edinburgh publisher, Robert Chambers (1802–71), who wrote (anonymously) the *Vestiges of the Natural History of Creation*, published in 1844. It was effectively a tract against the deism of William Paley's version of natural theology. Chambers wrote that when there is a choice between special creation and the operation of general laws instituted by the creator, 'I would say that the latter is generally preferable as it implies a far grander view of the divine power than the other.' Since there was nothing in the inorganic world 'which may not be accounted for by the agency of the ordinary forces of nature', why not consider 'the possibility of plants and animals having likewise been produced in a natural way'. The *Vestiges* was an instant best-seller. In the ten years following its publication, it sold more copies than did the *On the Origin of Species* fifteen years later. But it was full of errors. For Darwin, 'the prose was perfect, but the geology strikes me as bad & his zoology far worse'. Its importance was the debate it stirred. Darwin welcomed it on the grounds that 'it has done excellent service in calling in this country attention to the subject and in removing prejudices'.

The author of the *Vestiges* was widely attacked as a wild specu-lator, not to be taken seriously, but he signalled a warning to Darwin. Darwin's knowledge of the natural world was extensive, but it was not deep. Soon after the *Vestiges* appeared, Joseph Hooker commented in a letter to Darwin (September 1845) (*a propos* a pamphlet [1844] on the nature of species by a French botanist Frédéric Gérard, whom Hooker disparaged because he was 'neither a specific naturalist, nor a collector, not a traveller... and therefore a distorter of facts') that to be qualified to speculate about the nature of species:

> one must have handled hundreds of species with a view to distinguishing them & that over a great part – or brought from many parts – of the globe. I am not inclined to take much for granted from any one who treats the subject in this way and who does not know what it is to be a specific Naturalist himself.

Hooker presumably assumed that Darwin fulfilled these requirements, but Darwin was alarmed. He responded to Hooker, 'How painfully (to me) true is your remark that no one has the right to examine the question of species who has not minutely examined many.' He began a study of the barnacles he had collected on the *Beagle* cruise. This extended into a detailed investigation of barnacles worldwide and occupied him from 1846 to 1854; his barnacle work still remains a standard study of the group. But it became much more than a self-justifying exercise: it opened Darwin's eyes to variation within species and the apparent relationships of different families. When he returned to 'transformism', he was much more aware of biological complications in the real world than previously.

BOX 4.4   **Darwin's changing view**

The development of Darwin's ideas is well documented. His view of the natural world changed during his time on the *Beagle* as he saw the way animal species replaced one other along the length of South America and how fossils often resembled – but differed in details from – similar living forms. Notwithstanding, he still retained a 'traditional' belief in a world more or less as it was at its creation. The trigger that changed his views about this seems to have been his conversation in March 1837 with John Gould. Gould's finding that the finches living on the Galapagos were an entirely new group wholly confined to those islands forced Darwin to rethink his earlier assumption of an unchanging world, although he had no idea how this might have happened. The next year a mechanism which could change the status quo occurred to him when he read 'for amusement' Thomas Malthus's (1766–1834) *Essay on the Principles of Population*, which set out the spectre of the human population outstripping its food supply, condemning the weak and improvident to succumb in the consequent struggle for resources.

Darwin's genius was in linking the struggle for existence described by Malthus to the fact of heritable variation. If only a small proportion of a population survives the struggle, the likelihood will be that it will include those with a trait which gives its carriers some sort of advantage. Over the generations, the proportion of those with this trait would inevitably increase at the expense of those lacking it. There would be a genetic change in the population, amounting to 'natural selection' for the trait in question; natural selection leads to adaptation.

Twenty years after Darwin had read Humboldt in his Cambridge days, another 'amateur', Alfred Russel Wallace (1823–1913) also absorbed Humboldt, along with the *Voyage of the Beagle* and Chambers's *Vestiges of the Natural History of Creation*. Wallace had little money, initially working as a surveyor where he acquired a passion for insect collecting. He reacted to Humboldt in the same way as Darwin. It inspired him in 1848 to go to the Amazon with his entomologist friend Henry Bates (1825–92), beginning a trail which would lead ten years later to him writing to Darwin from the Moluccas in the Dutch East Indies (now Indonesia). Stricken with malaria, Wallace wrote a letter which showed he had come to the same conclusions about the mechanism of evolution as Darwin himself. It so horrified Darwin that it precipitated the announcement of evolution by natural selection at a meeting of the Linnean Society in 1858.

This public announcement was the spur for *On the Origin of Species*. It was written in a hurry at the instigation of Charles Lyell and Joseph Hooker. Darwin saw *On the Origin of Species* as little more than an abstract of the 'Big Book' he had been planning. He never wrote this. An unplanned consequence of Darwin's work was to drive biologists back into the field 'to watch Nature at work in her own way'.

On the Origin of Species was published on 1 November 1859 and sold out immediately. Darwin had steeled himself for reactions to his ideas. One criticism he probably did not anticipate was the claim that his 'dangerous idea' of natural selection was unoriginal. This complaint came from Patrick Matthew (1790–1874), an acerbic Scottish fruit farmer, who wrote to the *Gardeners' Chronicle* (an influential periodical at the time) to draw attention to his (Matthew's) priority in a book *Naval Timber and Arboriculture*, published in 1831. Darwin was relieved to discover that another Scot, William Wells (1757–1817), had suggested the same thing even earlier and in an even more obscure source – an *Essay on Dew*, published in 1818. Darwin was at first discomfited, but then reconciled by the consideration that the fuss was mainly about self-publicity on the part of Matthew. Encouraged by the Harvard botanist, Asa Gray (1810–88), Darwin added 'an historical sketch of the progress of opinion on the origin of species' to his 1861 revision of *On the Origin of Species*. He acknowledged a swarm of scholars who had preceded him in believing that species were subject to change into new species ('transformism' in the language of the time); he listed 34 authors, including Marchant (1719), Montesquieu (1721), Buffon (1749), Diderot (1764), his grandfather Erasmus Darwin (1794), St Hilaire (1795), Goethe (1795), Lamarck (1809), Grant (1826) and Baden Powell (1855), as well as Wells and Matthew.[14] But none of this should detract from the contribution and impact of Darwin himself.

At this point, Christian von Buch (1774–1853) enters the story. In his 'Historical Sketch' in *On the Origin of Species*, Darwin records, 'The celebrated geologist and naturalist, Von Buch, in his excellent *Description Physique des Isles Canaries* clearly expresses the belief that varieties change into permanent species, which are no longer capable of intercrossing.' This was made as an apparently throwaway comment in a description of an eight-month stay on the Canary Islands in 1815 with a Norwegian botanist, Christian Smith. Von Buch was primarily a geologist, and his contribution to evolutionary biology may have been unintentional. He had studied in Freiberg with Humboldt under Abraham Werner, and remained as friend of the former, often travelling with him. Notwithstanding, he seems to have been the first person to propose geographical speciation 'in a short statement which he failed to develop any further'. Nonetheless, Darwin recognized its significance.

Humboldt, von Buch and Darwin were fairly wealthy men, able to finance their travel without too much difficulty. Huxley had to survive on his salary as a naval officer. Wallace and his friend Henry Bates depended on selling specimens to well-to-do collectors. Another sort of contributor to knowledge and possible shaper of attitudes during the same period is represented by Richard Lowe (1802–74) and his friend, Thomas Vernon Wollaston (1822–78), both of whom

---

[14] Stott, R. (2012). *Darwin's Ghosts*. London: Bloomsbury.

studied under Darwin's teacher, John Henslow, in Cambridge. Lowe left Cambridge in 1827 to travel to Madeira for his health. With a few breaks, he stayed there, acting as chaplain to the English community from 1832 to 1854, until the British Home Secretary, Lord Palmerston, removed him from his post on account of his Tractarian extremism. He compiled a flora of Madeira. Wollaston stayed with Lowe several times. He was primarily an entomologist, fascinated by variation and species limits. Darwin and he corresponded, but Wollaston was unpersuaded by Darwin's transformism. He wrote a highly critical review of *On the Origin of Species*.

Lowe and Wollaston could be regarded as dedicated loners. This is a misleading distinction, because many of those who communicated widely and influenced future events and people began their involvement with the natural world simply because of their absorption and commitment to the world itself. John Ray, for example, 'was forced, following an illness that affected me both physically and mentally, to rest from more serious studies and to spend my time riding and walking, I had leisure in the course of my journeys to contemplate the varied beauty of plants and the cunning craftsmanship of Nature that was constantly before my eyes and had so often been thoughtlessly trodden underfoot'. This led him to write on the plants living around Cambridge, producing the first English local flora, published in 1660. He was effectively self-taught; he sought assistance in the University of Cambridge, but found no one to help him. A similar awakening came to Joseph Banks while still at school. He described his inspiration: 'walking leisurely along a lane, the sides of which were richly enamelled with flowers, [I] stopped and looked around, involuntarily exclaiming, How beautiful! After some reflection, [I] said to [my]self, it is surely more natural that I should be taught to know all these productions of Nature in preference to Greek and Latin...'

Then there are the parson-naturalists, who have contributed much to environmental understanding in Britain. The best known and most important of these is undoubtedly Gilbert White

(1720–93), for many years Anglican minister of the parish of Selborne in Hampshire. He wrote a series of letters to Thomas Pennant (1726–98), who may perhaps be best regarded as a pioneering proto-ecotourist, largely on the reputation of his guidebooks to highland Scotland – and who himself had been inspired with a passion for natural history through reading Ray's *Ornithology*, which he was given at the age of twelve. White became disenchanted with Pennant and published the letters he had written to Pennant (plus other material) in 1789 as *The Natural History of Selborne*. Years later Darwin wrote that 'From reading White's *Selborne* I took much pleasure in watching the habits of birds, and even made notes on the subject. In my simplicity I remember wondering why every gentleman did not become an ornithologist.' Another parson-naturalist who corresponded with Pennant was George Low (1747–95), a young clergyman from the Orkney Islands, who had begun work on a natural history of his native northern islands. Low collected information for Pennant on Orkney and the neighbouring Shetland group, but like Gilbert White before him fell out with Pennant, who took all the credit for Low's work for himself. Allegedly in large part due to this, Low died a disappointed man, aged forty-nine. The parson-naturalist tradition continued into the twentieth century, with practitioners like William Keble Martin (1877–1969), who wrote and illustrated with 1400 paintings a *Concise British Flora* (1965), which was a best-seller; Edward Armstrong (1900–78), an authority on bird behaviour and international expert on wrens (and author of three books of theology); and Charles Raven (1885–1964), Regius Professor of Divinity at Cambridge, biographer of John Ray and author of three books on bird-watching, plus a book of flower illustrations written with his son.

More typical 'loners' are John Muir (1838–1914), Henry Thoreau (1817–62) and Richard Mabey (1941–). Muir was a truly environmentally literate person. He had little formal education, but drew inspiration from journeys in the wilderness areas of North America, well expressed in his first book, *The Mountains of California* (1894)

FIGURE 4.5 John Muir (1838–1914), pioneer activist; patron of US environmental activity.
1907 photograph by Francis M. Fritz.

(Figure 4.5). Like Ruskin, he believed that 'everybody needs beauty as well as bread, places to play in and pray in, where nature may heal and give strength to body and soul'.

Muir's background and approach to writing about nature was not unlike that of Thoreau, whose account of experiencing life in a wooden cabin appeared in 1854 as *Walden*. Neither Thoreau nor Muir were scientists. Indeed, Muir was suspicious of the nascent ecology of his time, represented most strongly for him by Gifford Pinchot (1865–1914) and the power of the US Forest Service, of which Pinchot was the first chief. Muir and Pinchot fell out over the nature and management of wilderness. Muir valued nature for its spiritual and transcendental characteristics, and any management of it was sacrilege to him. Both Muir and Thoreau were inheritors of a growing tradition of wondering, learning and increasing understanding of the natural world that had been growing apace since the mid-eighteenth century. Richard Mabey is a more modern version of the same ilk, writing and broadcasting about the relationships between culture and the natural world. And perhaps David Attenborough ought to be included, although he studied biology at university and is a Fellow of the Royal Society. His contributions to shaping attitudes to the natural world as a broadcaster are enormous (p. 160).

Are these attitudes purely the result of a mindset produced by Western ideas? The Muslim scholar Seyyed Hossein Nasr has argued that Western civilization, in seeking to impose ever-increasing control of natural processes, has lost 'the sense of the spiritual significance of nature'.[15] For many people, the natural world has become merely something to be exploited. The US historian Lynn White argued much the same, in a very influential essay placing the blame squarely on Christian teaching (p. 236).[16]

All this is a massive change from the old view that the resources of the world are held by mankind on trust accountable to God for the use we make of them. A consequence is that in virtually all 'developed' parts of the world, we have become accustomed to central governments pervading almost all aspects of life. The state has effectively replaced many of the functions previously attributed to God. In terms of the management of resources, state intervention has meant two things: restricting individual freedom through planning restrictions, extending well beyond economic considerations to the maintenance of scenic beauty and the protection of wildlife; and through the amount of land held directly by government agencies, often linked to a concept of social benefit as a determinant of land use. These developments are generally regarded as 'progress' and can

[15] Nasr, S.H. (1968). *The Encounter of Man and Nature*. London: Allen Unwin
[16] The historical roots of our ecologic crisis. *Science*, **155**: 1203–7, 1967.

be justified as governments managing natural resources for the benefit of all, but they raise deeper questions about the 'value' of nature and the natural world. How should it be valued?

Values at the corporate and state level often conflict with those perceived by the individual. For example, a development seen as a community 'good' (such as a new road or airport) may diminish the worth of an individual's own property or infringe their privacy. A good example of this in Europe is the management and distortions of the Common Agricultural Policy – established with the best of motives to protect traditional farmers and farming methods but manipulated by national and commercial interests. The problem is that accounting procedures almost inevitably oversimplify valuations, usually by overweighting a few elements – most commonly by assuming that monetary considerations are all important. While acknowledging that such rationale is clear-cut and therefore tempting, it has to be acknowledged that it ignores much complexity and is potentially dangerous. Removing mangrove on tropical coasts may permit financially profitable enterprises such as the development of holiday resorts or prawn farming, but it exposes those who live nearby to coastal damage and, more devastatingly, to tsunami. The 2004 Indian Ocean tsunami swept up many unprotected shores; overall it killed almost a quarter of a million people. 'Filling in' spaces in towns may actually cause health problems to those who live there; moving from an urban to a rural situation seems permanently to improve 'well-being' (p. 198), including both physical and mental health.[17]

FURTHER READING

Armstrong, P. (2000). *The England Parson-Naturalist*. Leominster: Gracewing.

Browne, J. (1983). *The Secular Ark*. New Haven, CT: Yale University Press.

(1995). *Charles Darwin: vol. 1 Voyaging*. London: Jonathan Cape.

(2002). *Charles Darwin: vol. 2 The Power of Place*. London: Jonathan Cape.

---

[17] Shanahan, D.F. et al. (2015). The health benefits of urban nature: how much do we need? *Bioscience*, **65**: 476–85.

Darwin, C. (1887). *The life and letters of Charles Darwin, including an autobiographical chapter.* London: John Murray.

Holmes, R. (2008). *The Age of Wonder.* London: Harper.

Leroi, A.M. (2014). *The Lagoon: How Aristotle Invented Science.* New York: Viking.

McCalman, I. (2009). *Darwin's Armada.* New York: Simon & Schuster.

Mayr, E. (1982). *The Growth of Biological Thought.* Cambridge, MA: Harvard University Press.

Porter, R. (2000). *Enlightenment.* London: Allen Lane.

Raven, C.E. (1942). *John Ray, Naturalist.* Cambridge: Cambridge University Press.
   (1947). *English Naturalists from Neckham to Ray.* Cambridge: Cambridge University Press.

Stott, R. (2003). *Darwin and the Barnacle.* London: Faber.
   (2012). *Darwin's Ghosts.* London: Bloomsbury.

Wulff, A. (2015). *The Invention of Nature.* New York: Knopf.

5 Scientific Method and the New Biology – Controlling

The state may have been somewhat inadvertently taking over from God, but it has been a gradual process. Attitudes to the natural world at the beginning of the modern era, in Europe at least, were strongly influenced – virtually determined – by the interpretations of Thomas Aquinas (1225–74) and his version of natural theology, which was itself a combination of Aristotleanism and Christianity (p. 225). His vision was of a society guided by biblical tenets, one which firmly limited the ownership of property and condemned usury. The difficulty was that this seemed incompatible with economic development: either the rules had to be relaxed or economic development had to languish. As the Italian lawyer Benvenuto da Imola (1320–88) commented, 'He who takes usury goes to hell; he who does not, goes to the workhouse.'

The problem was that the Hebrew scriptures clearly restricted the charging of interest on loans. This had the positive effect of stopping unscrupulous entrepreneurs from exploiting the poor. In practice, such condemnation on interest was not absolute. John Calvin (1509–1564) argued that the underlying principle was that:

> the common society of the human race demands that we should not seek to grow rich by the loss of others... All unjust gains are ever displeasing to God, whatever colour we endeavour to give it... Usury is not now unlawful, except in so far as it contravenes equity and brotherly union. Let each one, then, place himself before God's judgement-seat, and not do his neighbour what he would not have done himself.[1]

---

[1] *Commentaries on the Four Last Books of Moses*, **4**: 125–33. English translation, Edinburgh: Calvin Translation Society, 1852.

A century after Calvin, John Locke (1632–1704) went further, effectively equating natural law with biblical revelation. It is said he 'invented common sense'. He regarded unused property as waste, and 'Though the water running in the fountain be every one's, yet who can doubt, but that in the pitcher is his only who drew it out? His labour hath taken it out of the hands of nature, where it was common, and belong'd equally to all her children, and hath thereby appropriated it to himself.'[2] In other words, Locke showed how to justify the accumulation of property over and above the requirements of its owner; everything changed once it was accepted that an individual's labour was owned by – or personal to –the labourer him (or her) self. It is said that his philosophy grew out of his laboratory work in Oxford on the physiology of respiration (inspired by his contacts with the chemist Robert Boyle), followed by his training and practice in medicine. They established for him the importance of observation and experiment, and the need to distinguish data from deduction. This carried over into his philosophical ideas. If the right to unlimited property is based on personal labour, property rights do not carry social obligations: society was not involved. This enabled Locke to overcome two objections: that someone should not own so much property that some of its products were wasted; and that the concentration of ownership reduces the resources available to the rest of the population. He answered the first of these by introducing the concept of money, which is not subject to spoiling, and the second by claiming that large-scale management increases overall production. These rationalizations meant that it became eccentric, if not quaint, to treat the 'goods' of creation as outside market forces.

All this was remote for most of the population. For them the environment was intensely insecure. Pain and sickness were commonplace. Life expectancy was low. Plague was endemic. The food supply was precarious; about one in six harvests failed completely, leading to deaths from starvation. These pressures were intensified

---

[2] Of property. In: *Second Treatise of Civil Government,* §29 (1690).

during the seventeenth and eighteenth centuries by rising commercial interests, which led to urbanization and new forms of exploitation of workers, together with the depopulation of rural areas. This was hastened by the enclosure of common lands. Enclosures meant that the traditional use of the commons for hunting, grazing and fuel collection were progressively denied to peasants, many of whom were forced to abandon their subsistence plots because they could not support a family without the associated commons. Enforced enclosures provoked widespread but ineffective protest. At the same time, transport and communication improved, enabling freer trade between towns and regions. Land and labour became counters in a money economy. The moral constraints imposed by the links between family and home were loosened. Historian Keith Thomas has described Tudor and Stuart England as 'an under-developed society, dependent upon the labours of an under-nourished and ignorant population'.[3]

The impact of such luminaries as Shakespeare, Milton, Locke, Wren and Newton was slow to filter down, but their influence did not exist in sealed bubbles. The Royal Society was formed in 1660 with specifically practical intentions (p. 120). Locke was an early Fellow. One of the Society's first initiatives (in 1662) was to set up a committee to enquire whether 'the cultivation of the potato would be a protection against famine'.

The attitude of the poor to their lot has been described as 'careless stoicism', but this did not render them immune to external events. The cascade of changes in society forced agricultural workers into rural landlessness or urban squalor, while fuelling the expansion of trade. It catalysed a transformation of feudalism towards capitalism. In his *Discourse on Inequality* (1754), Jean-Jacques Rousseau (1712–78) was explicit:

---

[3] Thomas, K. (1997). *Religion and the Decline of Magic.* London: Weidenfeld & Nicholson, p. 4.

> The first man who, having fenced in a piece of land, said 'This is mine', and found people naïve enough to believe him, that man was the true founder of civil society. From how many crimes, wars, and murders, from how many horrors and misfortunes might not any one have saved mankind, by pulling up the stakes, or filling up the ditch, and crying to his fellows: Beware of listening to this impostor; you are undone if you once forget that the fruits of the earth belong to us all, and the earth itself to nobody.

The biblical injunctions against usury and the accumulation of goods should have set the Church against these upheavals, but the Church itself as landowner and as economic actor was part of the emerging trading system.

BOX 5.1   **The tragedy of the commons**

William Forster Lloyd (1794–1852) was an Oxford University economist. He was critical of the optimism of some of his contemporaries about the market economy. One of his examples was the overgrazing of common land. He called this the 'tragedy of the commons'. Over a hundred years later a Californian ecologist, Garrett Hardin, developed Lloyd's example in a much-quoted article of the same name (*Science*, **162**: 1243–8, 1968). Hardin was concerned about the finiteness of resources in a finite world, complicated by the impending problem of overpopulation. His argument was simple. Imagine a common which can support forty beasts, with twenty farmers entitled to graze their animals. This means two beasts each. But any of the twenty may reason that there would be a negligible overall effect if they acquired a third animal; for the individual grazier there would be a spectacular 50 per cent increase in personal wealth at the expense of only one extra animal on the common. The problem is that all twenty might have the same thought, and sixty animals will appear on land capable of feeding only forty. Result: deterioration of both pasture and animals.

Hardin originally applied this principle to growing population pressures and the number of children any couple might agree to have, but later extended it to the way we treat the environment. For example, a manufacturer will be better off if he (or she) releases waste (sewage, chemicals or radionuclides) into a common outlet (river, air or sea), even though he (or she) then pays a 'share' of purifying the common. In such a case, voluntary cooperation for the group good would be largely fictitious. Hardin's reasoning has been criticized on the grounds that it only occurs if the use of the common in question is unregulated. Hardin was well aware of this problem and quoted Adam Smith (1723–90), the apostle of free trade and lack of regulation, who presumed that an individual who 'intends only his own gain... [is] led by an invisible hand to promote the public interest' (*The Wealth of Nations*, 1776). The problem was that (in Hardin's words), although 'Adam Smith did not assert that this was invariably true, he contributed to a dominant tendency of thought that has ever since interfered with positive action based on rational analysis, namely, the tendency to assume that decisions reached individually will, in fact, be the best decisions for an entire society.' In other words, personal ambition and greed can only too easily overcome altruism and the common good.

There are many instances where a tragedy of the commons has been worked out. One of the most spectacular was the fishery of the Grand Banks of Newfoundland, long seen as an apparently infinite source of cod. In the 1960s and 1970s, advances in fishing technology allowed huge catches. Following a few dramatically profitable seasons, the fish populations dropped, forcing fishermen to sail ever further to maintain catch sizes. By the 1990s, cod populations were so low that the Grand Banks fishing industry collapsed. It was too late for regulation and management; the cod stocks had been severely damaged and are only just beginning to recover. A similar example is the Antarctic whaling industry. Although it was clear to the whalers that the effort needed to catch whales was increasing enormously, indicating a decline in whale populations, the investment in boats and other plant meant that it continued until the early 1960s, despite being wholly uneconomic by then.

The principle of unfettered property rights is all very well, but it is dangerous. In Britain, the Companies' Act of 1862 made it possible for anyone to set up a limited liability company by simply obtaining signatures to a memorandum of registration. Prior to this, the ability to profit from commercial transactions was constrained because of liability for any obligations incurred. This check was removed by the Act. It became possible to grow rich without risk and in complete ignorance of business practices which would be outrageous if practised by an individual. It is said that 'the consequences of the Companies' Act of 1862 have been perhaps greater than any other measure in English Parliamentary history. It completed the divorce between the Christian conscience and the economic practice of everyday life. It paganised the commercial community.'[4] Another outcome has been an erosion of personal property rights towards a system of corporate social responsibility, with the state moving ever further into intervening in the economic and social life of its citizens. This has happened to different extents in different countries. In previous centuries there was a progressive shrinkage in the role of God and the Church; the situation now is where the state, concerned for its citizens, has widened its coverage to include personal welfare as well as economic issues.

This divorce of property rights from religious or social constraints has not gone unchallenged. The danger of limiting of self-sufficiency recurs in the writings of the nineteenth-century utilitarians. J.S. Mill was happy with the concept of private property, but believed that expediency should condition the ownership of land; access and enjoyment should be limited to that which is necessary for efficient exploitation. The ownership of land continues to inflame passion and cause conflict from Scotland to China, from Brazil to Zimbabwe.

Two positive notes need inserting into this litany of officially sanctioned management. The monastic system, particularly where

---

[4] Bryant, A. (1940). *English Saga (1840–1940)*. London: Collins, p. 215.

the Benedictine rule prevailed, provided a model of careful steward-ship throughout medieval Europe. Although various scandals marred the system from place to place, and its effect was much curtailed after the Reformation, the conscious care of creation encouraged successful environmental practices. The other development was a rationality arising from the scientific study of the natural world.

It is obviously wrong to claim that no rational appreciation of the environment existed before the modern era. Environmental disasters requiring response or amelioration have repeatedly afflicted humankind, dating back to the climate challenges of millions of years ago which, as it were, drove us down from the trees. Pollution must have been an issue from the early days of urbanization. Horace mentions the smoke which blackened buildings in Rome. Seneca was repeatedly advised to leave Rome for the sake of his health. He wrote to one Lucillus around AD 70 that no sooner did he leave Rome's oppressive fumes and cooking smells than he felt better. In Britain, Henry III's wife (Eleanor of Provence) moved 50 km from Nottingham to Tutbury Castle in 1253 because of the stench from coal burning in the town. Ironically, Mary, Queen of Scots, complained about the stink of the privies when she stayed in Tutbury in 1585. In 1306, the Knights Templar, owners of a mill at the mouth of the River Fleet (which opens into the Thames in central London), were prose-cuted for blocking the river and so preventing offensive offal from the butchers and leather-workers escaping into the main river.

In 1662 John Graunt, a draper, pioneered demography almost single-handed through an analysis he carried out by scrutinizing the weekly records of deaths kept by parish clerks (*Natural and Political Observations... Made upon the Bills of Mortality*). He showed that the death rate in London was much higher than in rural areas and argued that the explanation must be that this was due to the smoke-polluted air of London producing 'suffocations which many could not endure'. The diarist John Evelyn (1620–1706), who had written a tract on coal smoke called *Fumifugium or The Inconvenience of the Aer and the Smoke of London Dissipated*, seized upon this. In an oft-quoted passage he described London as shrouded in 'such a cloud of sea-coale, as if there be a resemblance of hell on earth, it is in this volcano in a foggy day: this pestilent smoak, which corrodes the very yron, and spoils all the moveables, leaving a soot on all things that it lights: and so fatally seizing on the lungs of the inhabitants, that cough and consumption spare no man'.

Evelyn laid blame for this squarely on the owners of a 'few Funnels and Issues, belonging to only Brewers, Diers, Lime-burners, Salt and Sope-boylers'. He saw no other reason for the air of London being so bad. The city had been built on a 'sweet and agreeable emi-nency of the ground', with a gently sloping aspect that allowed the Sun to clear the fumes from the waters and lower grounds to the south. Notwithstanding, it was another three centuries and many pea-soup fogs before the British Parliament passed a comprehensive Clean Air Act (1956). This has produced major benefits, but atmo-spheric pollution is still with us. It has been estimated that minute particles in the air from the use of solid fuel for heating and cooking and from diesel engines are responsible for over three million deaths a year.[5]

BOX 5.2   **Clean air legislation**

The control of atmospheric pollution is an environmental situation beyond the redress of most individuals, and hence needing regulation. The burning of coal was prohibited in London as long ago as 1273, because it was 'prejudicial to health'. Unsurprisingly, air pollution throughout the country worsened considerably as industry grew in the nineteenth century. The first modern restrictions in England were introduced in 1821, but they were largely ineffective, serving little more than to register that Parliament was against smoke. A major problem about legislating was that no witness could actually prove that smoke damaged health. Attempts were made to strengthen

---

[5] *Ambient (outdoor) air quality and health fact sheet.* Geneva: World Health Organization (http://www.who.int/mediacentre/factsheets/fs313/en/), 2016.

the law, but the chief effect of smoke pollution was indirect. Moreover, no one lived near polluting factories if they could avoid it – which meant being able to afford it. Those who had the money to move were also those able to influence legislation. This meant doing nothing. All this led to growing and lasting social segregation.

Graunt had shown a correlation between smoke and mortality, but correlation is no proof of causation – a point repeatedly made in more recent times in other contexts by apologists for the fossil fuel and tobacco industries. Endeavours to tighten smoke legislation in 1850 failed because (opponents claimed) it would cause unemployment, it was 'meddling with manufacturers' and it was premature in the current state of knowledge. Moreover, it had a low priority for the government in the face of trade depression, a cholera epidemic and famine in Ireland. However, in 1853 Home Secretary Lord Palmerston forced through a Smoke Nuisance Act, although this only affected London. It avoided the vested interests of the industrialists from the north, although it still attracted significant opposition. In discussion of the bill in parliament, Palmerston railed against those who opposed the Act:

> Here were a few, perhaps 100 gentlemen, connected with these different furnaces in London, who wish to make 2,000,000 of their fellow inhabitants swallow the smoke which they could not themselves consume, and who thereby helped to deface all our architectural monuments and to impose the greatest inconvenience and injury upon the lower class. Here were the prejudices and ignorance, the affected ignorance, of a small combination of men, set up against the material interest, the physical enjoyment, the health and comfort of upwards of 2,000,000 of their fellow men.

The problem got worse. In 1862 Lord Derby proposed the appointment of a select committee 'to inquire into the Injury resulting from noxious vapours evolved in certain manufacturing Processes, and into the State of the Law thereto'. The committee received evidence from three of the most distinguished chemists in the country (Lyon Playfair, August von Hofmann and Edward Frankland); it concluded that the 'monster nuisance' was hydrochloric acid released from alkali factories. Its report led to the setting up of an Alkali Inspectorate, with powers to enter factories if there was an environmental nuisance (as opposed to the perceived harm to workers, which was permitted under an existing Factory Act). This unintentionally marked the beginning of a scientific civil service, albeit with very limited powers. There was pressure to extend the remit of the Inspectorate, and a Royal Commission was set up which led to an increase in the nuisances regulated, most importantly in relation to cement and salt works.

By 1880, the problem was not so much industrial smoke in northern Britain as domestic smoke in the south. There were around three and a half million fireplaces in London, leading to thick and suffocating fogs which were almost as lethal as cholera. To deal with this involved tact and public relations rather than technical know-how. 'Clean-burning' anthracite stoves were available, but they were seen as no substitute for a 'cosy' fire. Ten attempts to legislate against domestic fires between 1888 and 1892 failed. The government was afraid to act. The Prime Minister, Lord Salisbury, excused himself:

> The difficulty is political... If this were to be made obligatory for Londoners, it would condemn the population to go for ever... without seeing a fire with a flame in it; I do not think that, for the sake of avoiding an occasional inconvenience, grave as it is, for a certain number of days in the winter, people would condemn themselves to a flameless fire all the winter through.

The Times wrote:

> We must endeavour to stir up public opinion to the point of action by implanting into the public mind that civilisation itself will be retarded by the toleration of nuisances that can be removed and of dirt that never ought to have been created... We are not without hopes that the day is not far distant when public opinion will insist that London must somehow be relieved of its canopy of gloom and its hideous vesture of dirt.[1]

The next decades were marked by some minor additions to the Alkali Act but also attempts to undermine official inertia. A Smoke Abatement Act in 1926 was very weak. Some cities passed their own

FIGURE 5.1A Atypical and melanic (*carbonaria*) peppered moths (*Amphydasis [Biston] betularia*) at rest on an unpolluted tree trunk in Dorset and on a smoke-blackened one just outside Birmingham. Reprinted by permission from Macmillan Publishers Ltd from Bernard Kettlewell (1956). Further selection experiments on industrial pollution in the Lepidoptera. *Heredity*, **10**: 287–301.

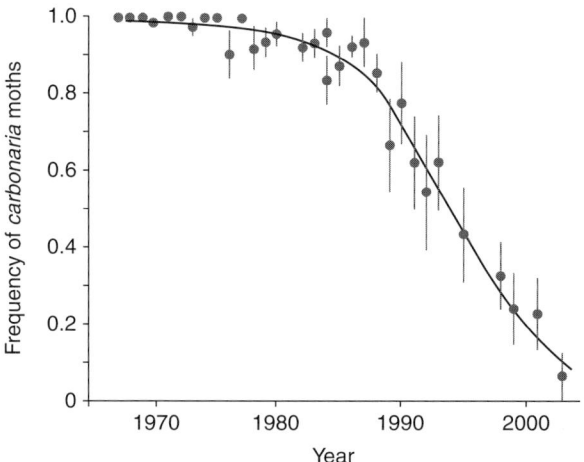

FIGURE 5.1B The decline in the frequency of *carbonaria* melanics following the revised Clean Air Act of 1964. The cleaner atmosphere allowed the regrowth of lichens on the moths' resting places, so providing camouflage for the typical peppered moths while exposing the melanic *carbonaria* to predation by insectivorous birds.
From John Thompson (2013). *Relentless Evolution*, by permission of University of Chicago Press.

smoke control orders, but there was no general control. The crunch
came in 1952 with a prolonged smog in London. Thirteen bulls
had to be slaughtered at the national beef show at the Smithfield
Market; an opera performance had to be halted after the first act
because the audience could not see the stage. Around 4000 premature
human deaths occurred. The government could no longer sit on the
fence. A comprehensive Clean Air Act came into effect in 1956. It
had various benefits beyond health protection. Smoke-blackened
buildings could be cleaned – and stay clean. A quantifiable gain
was the regrowth of lichens on trees. Most lichens are sensitive to
pollution. The disappearance of lichens from the mid-nineteenth
century reduced the camouflage of many species of moths in their
day-time resting sites, making them more visible and hence subject
to increased predation by birds. This led to the spread of genes which
replaced the old camouflage with black (or melanic) forms, which
were much less conspicuous to their enemies. Clean air reversed
this advantage: lichens returned and the old camouflage was needed
again. In both Britain and the USA the melanic gene in the most
studied species, the peppered moth (*Biston betularia*), has declined
markedly – from a frequency of 98 per cent in 1960 to less than 5 per
cent in a formerly highly polluted area (Manchester) (Figure 5.1).[2]

[1] *The Times*, 23 May 1890, p. 9.
[2] Cook, L.M. (2003). The rise and fall of the carbonaria form of the peppered
moth. *Quarterly Review of Biology*, **78**: 397–417.

BOX 5.3 **Acid rain and the ozone hole**

Smoke is not the only atmospheric pollutant. In the 1970s Norway
and Sweden became concerned about damage to their forests and lake
fisheries, which they attributed to long-range transport of sulphur
dioxide emissions from coal-burning power stations. Sulphur dioxide
could be removed from the smoke by 'scrubbers' in the chimneys,
but this was very expensive. The British Government claimed this
was unnecessary, as research showed that sulphur dioxide around
coal-burning power stations was very low if their discharges were

made from tall chimneys. However, ridicule was thrown on the British excuse by the discovery that most of the sulphur deposited in Norway came from elsewhere in Europe, and particularly from Britain. The government and the generating companies prevaricated; they argued that other factors than airborne pollution were involved (toxicity from aluminium, ozone, magnesium deficiency, drought) – quite apart from the fact that acidification of freshwaters had been increasing since the beginning of the Industrial Revolution. Then a major decline in foliage in German forests raised the stakes again. It seemed likely that air pollution was the main causative factor. The West German Government introduced a programme to halve sulphur emissions within ten years and to introduce catalytic converters to motor vehicles. This was followed by a similar commitment from France. Britain was branded 'the dirty man of Europe'. The government had little option than to authorize a large expenditure at coal-burning power stations to reduce the sulphur in emissions.

Although the underlying problem was very different, the ozone hole debates were not dissimilar in terms of the assertions and claims of interested parties. The hole was discovered over the Antarctic in 1984 by British Antarctic Survey scientists. Ozone (O_3) is formed naturally under the influence of ultraviolet (UV) radiation from the Sun by the addition of a third oxygen atom to normal oxygen molecules, which consist of two oxygen atoms (O_2). There is a layer of ozone in the atmosphere, 25–30 km above the Earth. This has the effect of absorbing UV radiation, which is a powerful carcinogen. Its importance is that it reduces the amount of UV reaching the Earth's surface. The layer thins every winter when it receives less sunlight, but no one expected that it would actually perforate.

It was known that chlorofluorocarbons (CFCs) widely used in refrigerators and as spray-can propellants (for deodorants, etc.) are active in breaking down ozone, and that they had increased significantly in the atmosphere. Banning CFCs should have been simple: it was a single group of chemicals with little doubt about their effects, made by a small number of manufacturers. But there was strong resistance, a reprise of the reaction to *Silent Spring* in 1962 (Box 9.1). The chairman of the chemical company DuPont

claimed 'the ozone depletion theory is a science fiction tale...
a load of rubbish... utter nonsense'. Nevertheless, considerable
international bargaining succeeded and an agreement to phase out
the use of CFCs (the Ottawa Protocol) was signed in 1987 (with
subsequent revisions). Even this was imperfect: the phasing out of
CFCs has been slower than it might have been – there are stories of
massive imports of CFCs into developing countries from countries
still making them, and there are concerns that the vast number
of fridges bought in some countries will continue to use CFCs
because they are cheaper than alternative 'environmentally friendly'
chemicals.[1] Notwithstanding, it is frequently hailed as an example of
what can be achieved by international agreement. Ozone depletion
is getting less, although it is not expected to attain pre-CFC levels
until 2050. For many years the Ottawa Protocol remained the only
universally ratified treaty in UN history.

The acid rain and the ozone hole debates both involved extended
periods of argument about the effects of sulphur dioxide and CFCs
respectively, needing further research and synthesis. Both resulted
in fairly acceptable agreements to mitigate their effects, but they
illustrate the complexity of dealing with environmental problems
where commercial and other vested interests are involved.

[1] Benton, T. (1994). *The Greening of Machiavelli*. London: Earthscan.

The eighteenth century is often described as the period of the
'Enlightenment', when society allegedly became more rational and
more 'scientific'. In principle, this gives tools for thought and action,
but these tools have to be used; they are not automatic agents. They
need moral commitment for their application. It is simply untrue to
claim that knowledge and understanding necessarily limit or automatically determine our actions. Notwithstanding, the Enlightenment
was a major step in our history. Historians debate whether it was
one great movement or whether there were distinct varieties, most
famously in France and Scotland, with local versions in England,
Germany and elsewhere. Some also distinguish between a moderate

Enlightenment and a radical one, strongest in France where it incorporated atheism, materialism and hedonism, never mind secret societies and Freemasonry. Whatever the details and however great the debates, there can be general agreement that the Enlightenment marked the decline (although it can never be said to be the end) of the fears and uncertainties at the end of the Middle Ages as portrayed by George Trevelyan (who has been described as 'a pious pagan who took his paganism with a deep reverence'):

> The idea of regular law guiding the universe was unfamiliar to the contemporaries of Francis Bacon [at the beginning of the seventeenth century]. The fields around town and hamlet were filled, as soon as the day-labourers left them, by goblins and will-o'-the-wisps; and the woods, as soon as the forester closed the door of his hut, became the haunt of fairies; the ghosts could be heard gibbering all night under the yew-tree of the churchyard; the witch, a well-known figure in the village, was in the pay of lovers whose mistresses were hard to win, and of gentleman-farmers whose cattle had sickened. If a criminal was detected and punished, the astonishing event was set down as God's revenge against murder; if a dry summer threatened the harvest, the parson was expected to draw down rain by prayer... The world was still a mystery, of which the wonder was not dispelled in foolish minds by a daily stream of facts and cheap explanations.[6]

BOX 5.4   **Profit versus safety: a Victorian cautionary tale**

Sadly, reason was still overridden by other interests. A highly discreditable situation about ship safety came to a head in the 1860s, well after rationalism was assumed to prevail. Ships were often overloaded and sank in conditions where they should have had no problems. The official regulatory body (the Board of Trade)

---

[6] *England Under the Stuarts.* London: Methuen & Co, 1904.

attributed at least half of the annual 1000 wrecks around the
British coast during the 1860s to overloading. In 1871 alone, 856
British ships sank within ten miles of the coast, many of them
in only a fresh breeze; another 149 went down in moderate gales
that should not have troubled sound and properly loaded vessels.
One in five merchant seamen drowned. Many of the ships that
foundered were unseaworthy 'coffin ships', bought cheaply at
auction as wrecks and then heavily insured by unscrupulous owners,
who profited enormously when the ship sank. When the Board of
Trade investigated, some ship-owners protested strongly against
'legislative interference' and 'restrictive practices which favour our
foreign rivals'. This self-serving argument even persuaded a craven
government to repeal a law against storing cargo on deck. The result
was ships were often top-heavy and even less stable.

All this led to pressure to regulate cargoes and stowage, led by
Samuel Plimsoll, a Member of Parliament inspired by the reforming
zeal of Lord Shaftesbury and the power of agitation which led to the
repeal of the Corn Laws (which kept the price of grain artificially
high in an attempt to help farmers). Regulation was opposed by a
small group of around eighteen ship-owners in parliament. Plimsoll
was accused of 'taking an exaggerated view of the facts'. He was
said to be acting out of self-interest because he had a business
transporting coal to London by rail, by-passing the sea-borne traffic
from Tyne to Thames. It was alleged by Oxfordshire MP Joseph
Henley that 'all this fuss about losses at sea is nothing more than
a "humanity dodge" to set up a monopoly for inland coal at the
expense of sea-borne coal'.[1] Plimsoll's response was a book, *Our
Seamen: an Appeal*, published in 1873. He was sued for libel – but
won. All this led to a Royal Commission on Unseaworthy Ships
being set up to examine the situation. In 1875 it concluded that
there was a need for safety standards for ships. The Commission's
recommendations were very weak, but a bill based on them was
accepted by the government. Then, to Plimsoll's horror, it was
dropped. This sparked public outrage; even Queen Victoria expressed
sympathy to Plimsoll. The government recanted. A Merchant
Shipping Act was passed in 1876. Clause 26 required that a mark or

'Plimsoll Line' be painted on the side of ships to indicate the limit of their safe load, allowing sufficient freeboard under different sea conditions.

[1] Jones, N. *The Plimsoll Sensation: The Great Campaign to Save Lives at Sea.* London: Little, Brown, 2006.

One way to describe the Enlightenment is by way of its key contributors: men like the thinkers and scientists already discussed, and to these we should perhaps add the influential romantic poets like Samuel Taylor Coleridge, Robert Southey and William Wordsworth. Another focus could be on the widening of human imagination through the awareness of an older Earth and an ever-larger cosmos. The enlargement of global consciousness from the travels of Cook, Banks, Darwin, Hooker, Lewis and Clark threw into sharper relief how different much of the world was to that of western Europe or eastern North America, not least from the descriptions and imports of these travellers themselves.

Widened perceptions must have come through the planting of non-native plants. Some of these imports had major impacts on society. Extracts from *Cinchona* tree bark were brought to Europe from Peru in 1648, providing quinine as a cure for malaria. *Cinchona* trees were successfully planted in Kew Gardens and taken from there to many parts of the world, giving protection against malaria to those with access to them. The profits from cultivating sugar cane in the West Indies fuelled the Atlantic slave trade. The East India Company used opium as payment for tea. Potatoes from South America became a staple food for industrial workers in the north of England – and agents of massive disruption when fungal blight led to the failure of potato crops. These and other such imports certainly influenced environmental attitudes in as far-reaching ways as the original domestications which led to the Neolithic revolution.

It was Joseph Banks who orchestrated the growing of *Cinchona* at Kew, and he certainly increased the popularity of plant introductions as a result of his time with James Cook (p. 82). But *Cinchona*, sugar, tea and spices were merely the more spectacular introductions: they were part of a tradition established much earlier. John Tradescant (1570–1638), and to a lesser extent his son, also called John (1608–62), brought fruit trees to Britain from the Low Countries for their employers, then journeyed to Russia and to north Africa, collecting many species of bulbs and seeds. They also acquired specimens from the US colonies. Hans Sloane (1660–1753) (whose collections were to become the basis of the British Museum) spent fifteen months in Jamaica as physician to the Governor. He brought back with him 800 species. Then a century later Banks returned in 1771 from his voyage with Captain Cook with 110 new plant genera and 1300 new species. When Cook refused to allow Banks to travel on his next voyage, Banks commissioned Frances Musson (1741–1805) as the first collector to be formally sent from the increasingly important Kew Gardens (effectively ruled by Banks). Musson spent his whole life collecting, first in southern Africa, then in the West Indies, finally in North America. He sent a constant stream of previously unknown species back to Britain. A few years later (1823), David Douglas (1799–1834) was sent to North America by the Horticultural Society (now the Royal Horticultural Society). He is most remembered for introducing the Douglas fir (*Pseudotsuga* spp.). Douglas was followed by a stream of collectors who financed themselves by selling specimens. Europeans became increasingly aware of and enthralled by exotic introductions. Rhododendrons first came to Britain in 1656. *Rhododendron ponticum*, which has gone on to become a major pest, arrived in 1753. The scale of introductions is shown by the fact that a third of the 4000 plant species which currently grow in Britain arrived after 1500.

More systematic study of the natural world was undertaken with the establishment in 1660 of the Royal Society of London, although its ambitions were explicitly utilitarian: 'to improve the knowledge of natural things and all useful arts, Manufactures, Mechanic practices, Engynes and Inventions by experiments'. A century later in 1788, the Linnean Society was founded, largely on the initiative of James Edward Smith (1759–1828), a Norfolk landowner who turned to medicine as a way to study botany while earning money. His fame largely rests on his purchase of Linnaeus's own collection of specimens and library from the latter's widow. The collections were initially offered to Joseph Banks but he did not want them, and advised Smith – at the time a young medical student – to acquire them. They cost him £1088. Smith apparently conceived the idea of a Linnean Society to provide a showpiece for his acquisitions – although he held on to them and after his death the Society had to buy them from his estate, crippling itself with the debt for many years. Smith saw 'his' society as complementing the Royal Society (of which he was a Fellow). His inaugural Presidential Address to the new Society stated that: 'It is altogether incompatible with the plan of the Royal Society, engaged as it is in all branches of philosophy, to enter into the minutiae of natural history; such an institution as ours is absolutely necessary to prevent all the pains and expense of collectors, all the experience of cultivators, all the remarks of real observers being lost to the world...' The Linnean Society is the oldest biological society still existing; its journal first published in 1791 is the oldest natural history periodical.

The emergence of the Linnean Society can be seen as a symptom of the growth of interest in the natural world towards the end of the Enlightenment period proper. It was founded in the same year that *The Natural History of Selborne* appeared. In one way it reflected the Enlightenment attitudes which allegedly accustomed people to view themselves and their surroundings with detachment rather than as a theatre for mere survival or enjoyment. In fact, the factors leading to this putative critical reality were not straightforward. In Scotland, James Hutton was emphasizing the impact of understandable processes of erosion and movement on the rocks beneath our feet and highlighting seemingly extraordinary realignments of land. 'Nature' came to represent a novel, albeit opaque mysteriousness. Landscape itself became 'picturesque'; William Gilpin's *Observations on the River Wye* (1783) introduced 'romanticism' into the language, encouraging viewing and writing about landscapes. Perhaps paradoxically, natural objects began to lose their power to overwhelm and even terrify, and became rather a reflection of striving and intuition. Collecting flowers, ferns, butterflies, birds and their eggs and stuffed animals of all sorts became increasingly common. As far as Britain was concerned, a number of influences can be seen as contributing to such passions: evangelical religion, middle-class earnestness, the absence of any dampening influence from professional science, easier publication through the introduction of steam-driven printing presses, the invention in 1896 of lithography and a lightening of the tax on paper.

BOX 5.5 **Romanticism and wilderness**

The Romantic Movement is conventionally linked to Rousseau (1712–78) in France and Goethe (1749–1832) in Germany. In Britain, William Gilpin (1724–1804) supported the notion of the picturesque as 'that kind of beauty which is agreeable in a picture'. For him, ruggedness (as in a tree or rocky mountain) differed from the beautiful. His *Observations on the River Wye and several parts of South Wales, etc.* (1782) was influential in stimulating travel and reacting against the rationalism of the Enlightenment. He was prescriptive about what makes a good view. He set out his own 'rules of picturesque beauty', commenting that 'If nature gets it wrong, I cannot help putting her right.' William Wordsworth visited the Wye Valley a few years later. It was he who really drove the Romantic Movement in Britain; its starting point is often taken as the publication in 1798 of *Lyrical Ballads* by Wordsworth and Coleridge. He even domesticated Isaac Newton, turning him (in his long poem *The Prelude*) from stern logician into 'a Mind forever voyaging through strange seas of Thought, alone'. He wrote a *Guide to the Lakes* (1810), turning on its head the judgement of Daniel Defoe in his *Tour thro' the Whole Island of Great Britain* (1724–7) that the landscape of the Lakes was 'the wildest, most barren and frightful of any that I have passed over in England, or even in Wales itself'. Perhaps to his chagrin, Wordsworth's fame attracted hordes of travellers to his beloved Lake District.

Without attempting to diminish the influence and contributions of these pioneers, it is proper to note they were anticipated in Britain by a growing emphasis on the 'natural'. A significant forerunner was Joseph Addison (1672–1719), not least through his editorship and writing in *The Spectator*, which he founded in collaboration with the traveller and wildly inventive Daniel Defoe.[1] Addison argued that the sublime brought together greatness, uncommonness and beauty; that it depended on wholeness, including mountains, deserts and seas, and its overwhelming impact was on the imagination. A key for him was 'greatness'. He wrote in *The Spectator* on 21 June 1712:

> By greatness I do not only mean the bulk of any single object but the largeness of a whole view considered as one entire piece… Our imagination loves to be filled with an object, or to grasp at anything that is too big for its capacity. We are flung into a pleasing astonishment at such unbounded views, and feel a delightful stillness and amazement in the soul at the apprehension of them.

A century later, John Ruskin (1819–1900) argued that the task of an artist was to be 'true to nature'. He denounced the idealized formality praised by his predecessors such as John Evelyn (who urged in his best-known book *Sylvia* [1664] that fruit trees be planted every 100 feet throughout the country, and that London be surrounded with enclosures of sweet-scented flowers), the way Thomas Gainsborough painted landscapes and the Italian landscapes of Claude Lorrain and Nicolas Poussin and, for that matter, Joseph Addison (who still influences us through his hymns, such as 'the spacious firmament on high… in Reason's ear they all rejoice'). Evelyn had written to Sir Thomas Browne that to him a garden was a restored arcadia, 'a princely and universall Elysium'. Ruskin regarded such an attitude as an abstraction derived from the landscapes described by Homer and made familiar in the eighteenth century through the translations of Alexander Pope. For Ruskin, such pleasant landscapes of cultivated fields and managed woodlands were distressing artefacts. They might be suited to human needs and convenience, but the danger was that

cultivation had become equated with beauty. He contrasted this
to the aesthetic of wilderness being espoused by Wordsworth and
his associates and which he himself advocated. For them 'nature'
was seen in the picturesque; mountains which had been previously
seen as 'the rubbish of the world' became part of the language of the
sublime (Figure 5.2).

Was this a watershed? For the historian George Trevelyan:

> The modern aesthetic taste for mountain form is connected with a moral
> and intellectual change that differentiates modern civilized man from
> civilized man of all previous ages... He now feels the desire and need for
> the greatness of untamed, aboriginal nature which his predecessors did
> not feel. One cause of this change is the victory that civilized man has
> now attained over nature through science, machinery and organization,
> a victory so complete that he is denaturalizing the lowland landscape...
> A new form of human desire has, under these conditions, arisen to get
> away from the vulgarity of man's triumph over nature, back to the old
> beginnings, to nature as God made her...[2]

FIGURE 5.2A Stylized attitudes. The 'picturesque' Wye Valley became
almost the quintessential example of a sublime landscape.
Photo © Matthew Lee Dixon.

FIGURE 5.2B Stylized attitudes. Ullswater in the English Lake District, complete with lake, hills, vegetation and solitude. This is where William Wordsworth 'wandered lonely as a cloud…'.
Photo © Drew Rawcliffe.

[1] Defoe's *Robinson Crusoe* has been variously read as an allegory for the development of civilization, as a manifesto of economic individualism and as an expression of European colonial desires, but it also shows the importance of repentance and illustrates the strength of Defoe's religious convictions. Robert Louis Stevenson claimed that the footprint scene in *Crusoe* was one of the four greatest in English literature.

[2] Trevelyan, G.M. (1931). 'The call and claims of natural beauty'. Rickman Godlee Lecture, delivered at University College London.

In his masterly account of *Books and Naturalists*, David Allen called the period at the end of the eighteenth century the 'Linnaean Spring'. He wrote:

All of a sudden around 1760… it became fashionable to go out into the countryside and try to put names to what one saw and found. This remarkable transformation had two separate causes. The more fundamental was the widespread awakening to the attractions of natural scenery in the layer of society that could afford to indulge in the leisurely tours now at last made feasible by marked improvements in the road system... Superimposed on that was a struggle to make sense of nature's almost overwhelming diversity, to work out the overarching plan on which it was assumed to be ordered... It was one more of nature's mysteries that was surely awaiting to be uncovered by the proper response of the intellect.[7]

White's *Selborne* was published in 1789, in the middle of all this. It focused the tradition. Samuel Taylor Coleridge annotated his copy, calling it 'a sweet delightful book'. It inspired Darwin to be an observer rather than a mere collector. In his *Autobiography* he wrote, 'From reading White's *Selborne* I took much pleasure in watching birds, and even made notes on the subject.' Virginia Woolf described the book as seeming 'to tell a plain story... and yet by some apparently unconscious device of the author's has left a door open, through which we hear distant sounds'.

The accuracy of the observations chronicled by Gilbert White was followed and complemented by the realistic and non-anthropomorphic animal paintings of George Stubbs (1724–1806), the engravings of Thomas Bewick (1753–1828) and the landscape paintings of John Constable (1776–1837), not least his passion for clouds, which paved the way for their scientific study. Together these contributed to a revolution in environmental perception which extended well beyond middle-class natural history pursuits to country dwellers in general (which was most of the population at the time). While this gentle revolution was being fomented among England's hedgerows, more canonical forms of revolution were gathering pace elsewhere. The French revolution occurred in the same year as *Selborne* appeared, 1789. Disorder was much less violent in Britain than in France and other European countries, and in the USA a few years later.

---

[7] *Books and Naturalists.* London: HarperCollins, 2010.

There was certainly considerable social unrest in Britain, including political volatility, economic depression, soaring prices and high unemployment. There was famine (p. 57), a surge of 'Luddite' machine smashing at the prospect of job losses, unemployment from the flood of ex-service men discharged after Waterloo and the end of the Napoleonic wars, and taxation at wartime levels. A reform meeting in 1819 in Manchester was charged by cavalry, resulting in the 'Peterloo Massacre' (although the massacre was small by most standards – only about a dozen people were killed). Ironically, the anarchy in France caused such a revulsion in Britain that it reduced the pressure for reform by the British Parliament. Continuing disorder was fairly low key. William Cobbett (1763–1835) railed against the inequalities and injustice experienced by farm workers, and his agitation was one of the factors which eventually led to the passing of a Reform Act in 1832.

The growing availability of printed material undoubtedly encouraged interest in the natural world. Unwieldy by modern standards, but at the time a notable production, was an *Introduction to Entomology* in four volumes (1815–26) by William Kirby, an Anglican clergyman and correspondent of Linnean Society founder James Edward Smith (and twenty years later an author of one of the Bridgewater Treatises – see p. 86) and William Spence, a Yorkshire businessman. A serendipitous boost to field botany was given by the Apothecaries' Act of 1815, which required all medical students to have some knowledge of botany, since most drugs in use were derived from plants. This led to regular field classes being organized. These in turn stimulated an enthusiasm for field collections. Natural history societies sprang up all round Britain, most of them with a local focus and membership. The dedication and enthusiasm of some of their members in building of collections of pressed specimens led to the exchanging of surplus specimens to produce more complete collections. The existence of collections of species from different areas evoked the possibility of using the information in them to investigate the distribution of species, permitting finding out more about the causes behind these distributions. A 'Botanical Exchange Club' came into being, initially as a small enterprise in Yorkshire but spreading to become national, eventually becoming the Botanical Society of the British Isles.

At the British Association for the Advancement of Science meeting in 1904, Arthur Tansley, a young botanist on the staff of University College London, proposed a 'central committee' to survey and map the British vegetation. The committee was duly established with Tansley as secretary, and laboured profitably. In 1912, its members became the founding council of the British Ecological Society, the first such society in the world.

This enthusiasm for the natural world in the early part of the nineteenth century was nurtured by many popular books. One book by the Rev John G. Wood (1827–89), *Common Objects of the Country* published in 1859, is said to have sold 100,000 copies in a week. The Education Act of 1870 required elementary science to be taught in schools – which meant that teachers had to be trained. But the moves which led to the formation of the British Ecological Society were also signs of an increasing professionalization of science. Michael Foster, President of the Yorkshire Naturalists' Union and Secretary of the Royal Society, welcomed the impact of Darwinism which had taken the biologist back 'into the field to watch Nature at work in her own way', leading to the rehabilitation of the naturalist of old. Notwithstanding, he mourned 'the whole-minded inquirer who had been cut into little bits, and while such bits as the study of form, structure, function, habits and history had flourished and grown great, the whole had vanished from sight'. It was a period when the laboratory biologist or physiologist became a hero, looking down on the luckless being, content to search for species.

Robert Lloyd Praeger (1865–1953) reacted similarly. He was the only member of the 'Central Vegetation Mapping Committee' who was not a college lecturer; he was a very accomplished and energetic field botanist, but his job was as a university librarian. In 1922 Praeger became President of the British Ecological Society and wrote how:

the question 'why' kept intruding itself, becoming more insistent and clamorous as time went on… So it came about that the glorious days of primary survey, when we ranged free over moor and mountain, to a great extent were superseded. Our campaign took on a new phase, and weapons of greater accuracy were required. Six-inch maps, binoculars and pencil were replaced or at best reinforced by instruments for measuring the amount and variation of light, heat, moisture, and the whole battery of the chemical laboratory.[8]

The appearance of more and more specialized journals had the effect of diluting even more the contributions of the local societies to the corpus of knowledge.

Charles Kingsley (1819–75), best known as the author of the *Water Babies* but also Regius Professor of Modern History at Cambridge, wrote that in his youth the naturalist had been regarded as a figure of fun, a 'harmless enthusiast who went bug-hunting simply because he had not the spirit to follow a fox'.[9] He believed that the trigger for the spread of environmental interest was the *Natural History of Selborne*. It was usually a fairly unsophisticated enthusiasm. Popular writers routinely advocated a crude Paleyan natural theology, quoting Pope's *Essay on Man* (concerning the one 'who takes no private road, but looks through Nature to Nature's God', a concept taken up by Thomas Jefferson in the Declaration of Independence) and Shakespeare in *As You Like It* (concerning the one who 'finds tongues in trees, books in the running brooks, sermons in stones'). In other words, the primary reason to study the natural world was to find that God exists, the second to illustrate His attributes. Natural history writing was often pietistic and moralistic. In a *Notebook of a Naturalist* (1845), E.P. Thomson applauded 'an all-wise and beneficent Providence to assign birds to free us from the clouds of insects, which would otherwise infest our dwellings

---

[8] Annual Meeting, 1922. *Journal of Ecology*, 11(1): 112–23, 1923.
[9] The Wonders of the Shore. *North British Review*, **22**(43), 2–56, 1854.

and destroy the labours of the field... The day has happily passed in which the votaries of nature were taunted with ridicule and addicted to childish fancies.'

In retrospect, all this seems extremely naïve, but there is ample literary evidence to show how common it was. Alfred Tennyson wrote about 'nature, red in tooth and claw' in his long poem *In Memoriam* published in 1850. William Kirby, the author of the influential *Introduction to Entomology*, wrote in a letter in 1800, 'The author of Scripture is also the author of Nature: and this visible world, by types and by symbols, declares the same truths as the Bible does by words.' He also wrote 'The larva of the Ichneumon, though every day, perhaps for months, it gnaws the inside of the caterpillar, and though at last it has devoured almost every part of it except the skin and intestines, carefully all this time it avoids injuring the vital organs, as if aware that its own existence depends on that of the insect upon which it preys.'

Darwin was well aware of this habit, although he did not accept Kirby's teleology. In 1860, the year after the publication of *On the Origin of Species*, he owned up to his Harvard friend Asa Gray:

> I own that I cannot see as plainly as others do, and as I should wish to do, evidence of design and beneficence on all sides of us. There seems to me too much misery in the world. I cannot persuade myself that a beneficent and omnipotent God would have designedly created the Ichneumonidae with the express intention of their feeding within the living bodies of caterpillars, or that a cat should play with mice.

FURTHER READING

Allen, D.E. (1976). *The Naturalist in Britain: a Social History*. London: Allen Lane.
Ashby, E. and Anderson, M. (1981). *The Politics of Clean Air*. Oxford: Clarendon Press.
Barber, L. (1980). *The Heyday of Natural History*. London: Jonathan Cape.
Bate, J. (2000). *The Song of the Earth*. Cambridge, MA: Harvard University Press.

Crane, N. (2007). *Great British Journeys*. London: Weidenfeld & Nicolson.

Harman, P.M. (2009). *The Culture of Nature in Britain 1680–1860*. New Haven, CT: Yale University Press.

Hobhouse, H. (1985). *Seeds of Change*. London: Sidgwick & Jackson.

Livingstone, D. (2008). *Adam's Ancestors: Race, Religion and the Politics of Human Origins*. Baltimore, MD: Johns Hopkins University Press.

Pavord, A. (2016). *Landskipping*. London: Bloomsbury.

Porter, R. (2000). *Enlightenment*. London: Allen Lane.

Rudwick, M.J.S. (2014). *Earth's Deep History*. Chicago, IL: University of Chicago Press.

6 Science in Public Affairs – Organizing

Democratically elected governments are expected to respond to the needs, aspirations and challenges of their subjects. Some of their actions may have more far-reaching effects than their initial purposes, however. One of the clearest manifestations of this has been the provision of maps. Their original purpose was for the military but the production of increasingly accurate maps helped those living in towns to identify markets and routes which stimulated the building of road links, and this in turn encouraged travel.

Map-making in Britain was provoked by the frustrations of the military needing to move troops quickly during the Jacobite rebellions, together with chagrin about the failure of the Navy to capture Bonny Prince Charlie after Culloden. In 1747 a young Scotsman, William Roy (1726–90), was given the job of mapping the Highlands. He was so successful that his remit was extended to the whole of Scotland in 1752. By 1755 he and his assistants had working drafts for the whole of that country. Then Roy was moved south, to take part in a reconnaissance of the invasion-sensitive coast from Dover to Milford Haven. The Seven Years War (1756–63) and the risk of French invasion in the south had led to a need for good maps of southern England. By 1766 Roy was petitioning the king about the need for 'a General Military Map of England'.

The next development came in 1783 with an approach by a group of French astronomers to Joseph Banks as President of the Royal Society to instigate a measurement of the distance from Paris to Greenwich; they believed that the latitude of Greenwich was wrong. Linking the observatories in the two cities would put global navigation on a firmer foundation. Banks saw the French request as 'doing honour to our scientific character'. He passed the French note to Roy. The French had been mapping their own country for some years, using the method of triangulation. The king provided £2000 for Roy to begin 'an accurate survey of the British Dominions'. This got under way in 1784. A line measuring 5.19 miles (8.35 km) was laid down on Hounslow Heath, where Heathrow Airport now stands. It was overseen by Roy and watched by a host of notables, including Joseph Banks. This line was and is the base for the triangulation mapping of the whole of the United Kingdom. The Ordnance Survey was born, an agent of the military. The triangulation was hard work; the primary work lasted until 1841. A series of one inch to the mile maps for the whole of Great Britain was only completed in 1891. British maps are now among the best in the world, although their military use within the islands must be negligible.

BOX 6.1   **Ocean navigation**

From earliest times, seafarers would have passed on their knowledge of local hazards and landmarks in their familiar inshore waters, although the charts they produced were usually sketchy and rudimentary. Notwithstanding, reasonably accurate charts of the Mediterranean coasts were available by Roman times. The first accurate charts of the British coast were of the Orkney Islands, compiled by a local man, Murdoch Mackenzie (1712–97) in the 1750s. Their value led to his employment by the Admiralty to survey more western coasts.

Orienting on land (or near the coast) with visible landmarks is one thing. Finding one's position in mid-ocean is very different – or at least it was before the days of positioning systems based on signals from radio beacons and orbiting satellites.

It is necessary to know both latitude and longitude when navigating away from land. Latitude is fixed by nature; it can be determined by the length of the day, the height of the Sun or knowledge of guide stars. Longitude depends on time. The twenty-four hours of each day represent the Earth rotating 360 degrees – fifteen degrees for each hour. Before the days of positioning systems, sailors were forced to depend on 'dead reckoning', knowing how far they had travelled over a period of time. Accurate determination of longitude mean could only come from knowing the difference in time from some point of known longitude. The Spanish and then the Netherlands Governments offered prizes for anyone who could devise a method for determining the longitude of ships at sea.

By the mid-eighteenth century the growth of trade made the need for maps of offshore waters increasingly acute. Captain James Cook was employed by the British Admiralty as a result of his chart-making skills learnt mapping the St Lawrence River in the Seven Years War. It was because of his efficiency and expertise in Canada that he was then sent to the Far East to explore the Southern Ocean. Half a century later, another dedicated chart-maker, Robert Fitzroy, was despatched to map the shoreline of South America, in command of HMS *Beagle*.

A crisis for Britain came on 22 October 1707, when four out of a fleet of five British warships twelve days out from Gibraltar were wrecked on the Scilly Isles, with the loss of over 1550 sailors. The British admiral Sir Cloudesley Shovell was confident that he was clear of the coast of Brittany, but misjudged his position catastrophically. It was one of the worst maritime disasters ever to afflict Britain. The consternation that followed led to the establishment in 1714 of a Board of Longitude with the then massive sum of £20,000 offered for any 'practicable and useful' solution to measuring longitude. The Board squabbled long and with discredit over possible solutions, but eventually (and only after the personal intervention of the king, George III, in 1773) awarded the money to a self-taught clockmaker, John Harrison, for inventing an accurate timekeeper – a chronometer capable of being carried on board ship. To qualify for the award meant that the watch must not lose or gain more than three seconds in twenty-four hours. A trans-Atlantic voyage showed Harrison's chronometer losing a paltry five seconds after eighty-one days at sea.

The Ordnance Survey was originally paid for by military funds. There was no mechanism for government funds to be generally available for such purposes. Notwithstanding, the practice of ad hoc grants for specific scientific projects seems to have become commoner around 1830. Charles Darwin was given £1000 towards publishing the results from his *Beagle* voyage, following a personal interview with the Chancellor of the Exchequer. He was later successful in a petition for a pension for Alfred Russel Wallace.

The 1830s were an age of 'reform' in the political world but also a time of change in all sorts of other arenas. Science was becoming professionalized; with it, there was a growth of institutionalized influencing of attitudes. In 1828, a rather acerbic mathematician, Charles Babbage (Lucasian Professor of Mathematics at Cambridge; best known for inventing the first programmable computer), attended the annual meeting of the Society of German Scientists (Gesellschaft Deutscher Naturoforscher und Ärzte), founded in 1822 by Lorenz Oken to provide a forum for discussion in medicine and natural science. Babbage was the only Englishman there. He was depressed about the state of science in Britain, and obviously impressed by the meeting. He speculated about a larger European gathering. In 1830, he published *Reflections on the Decline of Science in England and Some of Its Causes*, a vitriolic and multipronged attack on the lack of government support for science, the neglect of science in the universities, Britain's loss of scientific pre-eminence to continental researchers and the incompetent governance of the venerable and entrenched Royal Society. It spawned a 'Declinist Movement', bemoaning the fact that lack of support for British science was harming British interests.[1] Babbage argued that material prosperity required pure research. The geologist Charles Lyell supported Babbage. Edinburgh-based physicist Sir David Brewster, the editor of the influential *Quarterly Review*, wrote positively about Babbage's ideas. Brewster and Babbage had a vested interest in their attack; they

---

[1] Hall, M.B. (1984). *All Scientists Now*. Cambridge: Cambridge University Press.

were increasingly dependent on institutional support for the large and more complex instruments they wanted to create. In the same year Samuel Taylor Coleridge produced *On the Constitution of the Church and State, According to the Idea of Each*, arguing that the morality of any state required religion for its continued health. He was worried about moves to popularize science.

Meanwhile, various provincial 'philosophical' institutions had come into being – in York, Manchester, Liverpool, Bristol. 'Philosophical' in this context meant activities and studies stretching over both social and natural sciences. Attempts were made to reform the Royal Society, but in vain; Brewster and Babbage became more and more frustrated. Brewster decided that 'the cause of science in England would derive great benefit from a meeting of British men [*sic*] of science in York… as the most centrical [*sic*] city for the three Kingdoms'. Their incitement led to the formation of a British Association for the Advancement of Science, which held its inaugural meeting in York in 1831 in the face of disinterest and positive opposition from London and Oxbridge. Partly in response to this, a second meeting was held in Oxford in 1832 (dismissed by *The Times* as 'a mere unexplained display of philosophical toys'), but the organization finally came to national acceptance the next year at its third meeting in Cambridge. Edinburgh followed in 1834, then Dublin in 1835. The term 'scientist' was coined at the Cambridge meeting by William Whewell, Master of Trinity College, to recognize that scientific knowledge was different from other forms of knowledge; it did not seem particularly meaningful to call those who practised science 'natural philosophers'.

The British Association provided a visible focus for public support for science. Almost from the beginning, it gave grants to individual scientists. In addition, and at its behest, an annual government grant of £1000 was given to the Royal Society from 1850, increased to £5000 a year from 1877. It had wholly unexpected consequences. Although the French and German Governments had long given grants to support science, and the British Association had supported specific projects, this was the first time in Britain that money was provided for unspecified scientific purposes. At the time there were very few paid posts in science; the distinction between amateur and professional was still in the future. The physiologist W.B. Carpenter wrote:

> If England was behind Germany in original investigations, it was not, as is sometimes said, because Englishmen are inferior to Germans in ideal power, but because the German universities are so arranged to afford a career to men who choose to devote their lives to study. In England such men, having no means of making a livelihood by the pursuit of science, are obliged to turn their attention to a 'practical profession'.[2]

One biologist of the time advised his pupils to give up their study in favour of chemistry since 'there was no pursuit of a career for them in anything else'.

There is no doubt that the existence and influence of the British Association (now the British Science Association) catalysed the professionalization of science. A new profession of 'science' grew up, mainly in the universities. Although there was no formal apartheid between professional scientists and the large number of 'amateurs' – some of the latter were highly knowledgeable and consorted regularly with the 'professionals' – the increased number of specialist periodicals and their technical language reduced the ability of general natural historians to communicate with those employed in the universities.

BOX 6.2   **1859 and all that**

The publication of *On the Origin of Species* in 1859 was a tipping point for science in general and biology in particular. It meant that attitudes to the natural world had to change, never mind the need to re-evaluate humankind as an integral part of the natural world. The fact of evolutionary change was rapidly accepted but other re-focusings were slow in coming. Evolution made sense of so many

---

[2] University organization. *The Academy*, 3(61): 459–60, 1872.

data – of comparative anatomy and physiology, of classification, of geographical distribution, of fossil relationships. It received an effective *imprimatur* when Frederick Temple (soon to become Archbishop of Canterbury) echoed Charles Kingsley, who had proclaimed, 'We knew of old that God was so wise that he could make all things: but behold, He is so much wiser than even that, that he could make all things make themselves.'[1] In his 1884 Bampton Lectures, Temple agreed: '[God] did not make the things, we may say no: but He made them make themselves.' Ironically in the light of future history, Darwin's ideas were assimilated more readily by conservative theologians than by liberals, apparently because of the stronger doctrine of providence of the former. Some of the authors of the 'Fundamentals', a series of booklets produced between 1910 and 1915 to expound the 'fundamental beliefs' of Protestant theology as defined by the General Assembly of the American Presbyterian Church and which introduced the word 'fundamentalism' into the language, were sympathetic to evolution. Princeton theologian B.B. Warfield, a passionate advocate of the inerrancy of the Bible, believed that evolution could provide a tenable 'theory of the method of divine providence in the creation of mankind'.[2]

In contrast, what Darwin regarded as his 'big idea' of natural selection as the main mechanism of evolutionary change was less successful. It did not fit easily with conventional notions of progress or improvement, and philosophers and theologians developed alternative proposals – albeit more metaphysical than scientific. Even worse, the rediscovery of Mendel's results in 1900 and the subsequent explosion of the science of genetics seemed to show that the physical basis of heredity based on genes and chromosomes could not be the basis of the sort of variation expected by Darwinism. Mutations studied in the laboratory tended to have a large effect, to be deleterious and to be inherited as recessive traits. The Darwinian expectation was that evolution progressed through small steps produced by favourable variants. In a book surveying the state of evolutionary biology published in 1907 in preparation for the Jubilee of *On the Origin of Species*, a distinguished US entomologist, Vernon Kellogg, began by proclaiming 'the death-bed of Darwinism'.

It took two or three decades more before the work of R.A. Fisher, J.B.S. Haldane and Sewall Wright in the 1920s brought genetics, palaeontology and comparative studies together in a 'neo-Darwinian synthesis'; this rehabilitated Darwin's original thesis and has proved resilient in the face of further discoveries – notably the challenges produced by the 'molecular revolution' of the 1960s and 1970s. But these debates are a side issue for most people. (Outside the scientific community proper, the only group to take much interest were 'creationists' eager to find holes in the evolutionary account.)

[1] The natural theology of the future. In: *Westminster Sermons*. London: Macmillan, 1874.

[2] On the antiquity and unity of the human race. *Princeton Theological Review*, **9**:1–25, 1911.

The British Association for the Advancement of Science welcomed both amateurs and professionals, recognizing that the lack of financial support for science had spread ripples because of the lack of jobs for scientists. The universities had to react. In 1850, Royal Commissions were set up to inquire into Oxford and Cambridge (the only universities in England at the time). The same year, Oxford introduced an Honours School of Natural Sciences; the following year Cambridge set up a Natural Sciences Tripos. These developments were resisted by some, particularly in Oxford where there were protests about the 'Germanizing' of education. But by 1860, it was clear that these responses by Oxbridge were not enough. British manufacturers failed to shine in the industrial section of the Paris International Exhibition in 1867. Teaching needed to be complemented and supported by research. The poet Matthew Arnold, son of the reforming headmaster of Rugby School, placed the blame for British failures on the inadequacy of secondary education in a powerful diatribe on *Schools and Universities on the Continent* (1868).

These rumblings and unease had powerful support at the heart of what would now be called 'the scientific establishment': the X Club (Figure 6.1), an unofficial dining club of nine members, initiated

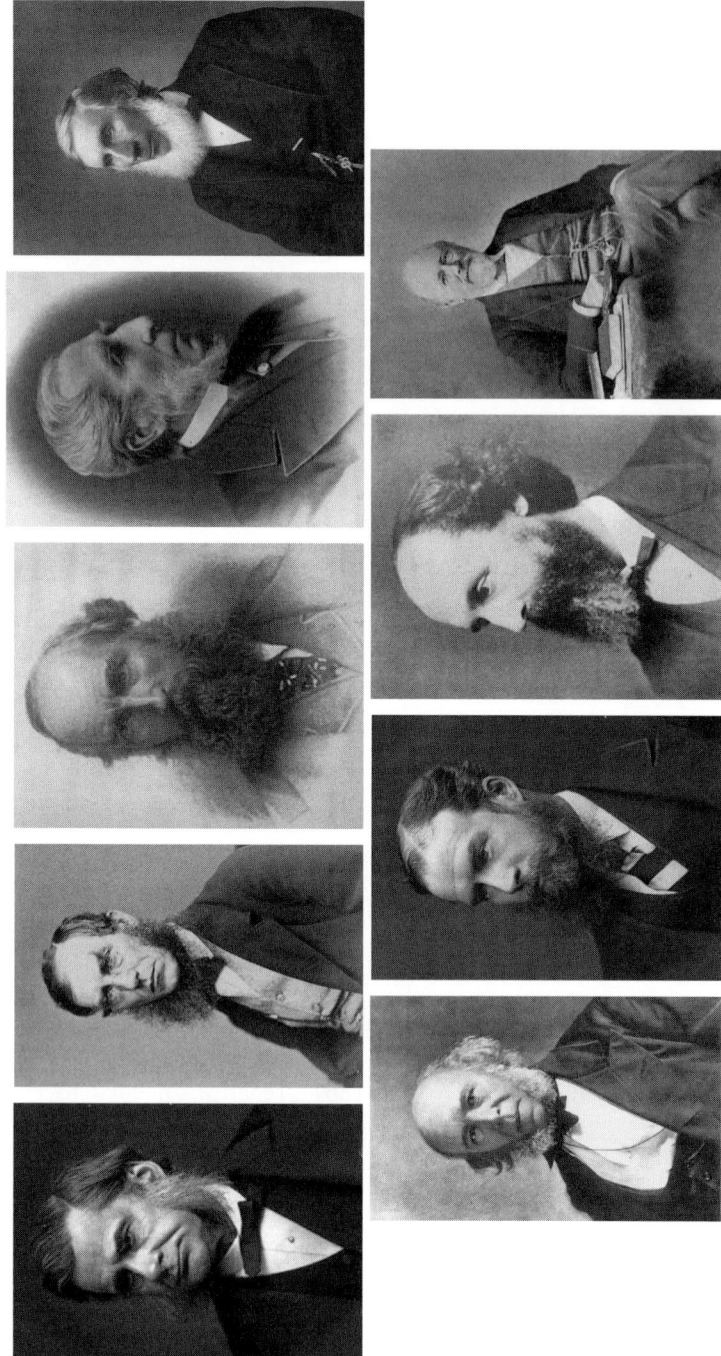

FIGURE 6.1 The X met monthly from 1864 to 1893. (upper row from left): Thomas Huxley (zoologist), Joseph Hooker (botanist), William Spottiswoode (mathematician), George Busk (surgeon and anatomist), John Tyndall (physicist); (lower row): Herbert Spencer (sociologist), John Lubbock (ethnologist), Thomas Archer Hirst (mathematician), Edward Frankland (chemist).

by Thomas Henry Huxley with the strong support of Darwin's confidant, Joseph Hooker, Director of the Royal Botanic Gardens at Kew. The X Club acted as a pressure group to influence government and society in their attitudes and support for science.

The X Club differed from the 'Declinists' because its members were at the heart of trends in scientific life and of education outside Britain. They were also conscious of their own financial struggles. The X Club met almost monthly from 1864 until 1893, taking up the baton of Babbage and Baxter and the Declinists with the aim of reforming the Royal Society and making science more professional. Both Huxley and Hooker subsequently served as President of the Royal Society (as did another member, the mathematician William Spottiswoode). They were united by a 'devotion to science, pure and free, untrammelled by religious dogma', although they were not antagonistic or disinterested in religion. They backed the theological collection *Essays and Reviews*, published in 1860, which caused a greater stir than *On the Origin of Species*, published the previous year.

BOX 6.3 Thomas Henry Huxley

Huxley (1825–95) is a key figure in the professionalization and secularization of science in the second half of the nineteenth century. He served on eight Royal Commissions, was president of the Royal Society (1883–5), popularizer of Darwin's evolutionary ideas (although he never seems to have fully understood or accepted the importance of natural selection) and massively influenced science teaching at all levels.

Huxley coined the term 'agnostic' to describe a person who holds that we cannot know anything of things beyond material phenomena, allegedly basing it on a sermon of St Paul reported in the Acts of the Apostles (chapter 17). He was vehemently opposed to superstition and unsupported claims, but it would be wrong to characterize him as an atheist. Towards the end of his life in 1892 he wrote in an unpublished essay, 'An Apologetic Eirenicon':

> It is the secret of the superiority of the best theological teachers to the majority of their opponents that they substantially recognize the reality of things, however strange the forms in which they clothe their conceptions. The doctrines of predestination, of original sin, of the innate depravity of man and the evil fate of the greater part of the race, of the primacy of Satan in this world, of the essential vileness of matter, of a malevolent Demiurge subordinate to a benevolent Almighty who has only lately revealed himself, faulty as they are, appear to me to be vastly nearer the truth than the 'liberal' popular illusions that babies are all born good and that the example of a corrupt society is responsible for their failure to remain so.

John Lubbock, one of the X Club members (and a near neighbour of Charles Darwin in Kent) even suggested an alliance between liberal Anglicans and scientists. Distinguished scholars who were invited to dine included August Laurel, Hermann Helmholz and Alfred Cornu from continental Europe; and Asa Gray, Louis Agassiz and E.L. Youmans from the USA.

Members of the X Club shared the conviction of the Declinists that science needed and was worth public support. To a scientist without a private income the options were limited – school teaching, emigration or protest. Both the physicist John Tyndall and the chemist Lyon Playfair were only prevented from seeking jobs in Australia or North America by Prince Albert and Prime Minister Robert Peel contriving posts for them (at the Royal Institution and the School of Mines, respectively). In 1868, the British Association for the Advancement of Science set up a committee composed mainly of X Club members and chaired by Playfair to examine whether there was sufficient provision for physical science in Britain.

The Playfair committee argued that teaching and research were interdependent. Its advocacy produced quick results, with the setting up in 1870 of a Royal Commission on 'Scientific Instruction and the Advancement of Science', chaired by the Duke of Devonshire. The Commission sat for six years, met eighty-five times, interviewed over

150 witnesses and published eight reports. Its third report (1873) concluded that 'the lack of pecuniary means can be the main difficulty which has hitherto in the richest country of the world, hindered original investigation in the sciences. The natural harvest which scientific discoveries in England might annually reap has... been checked by the irregularity with which the labourers have been rewarded and the comparative indignity with which they have been treated.'

The lack of research facilities in Britain led to a generation of scientists who were forced to work in laboratories on continental Europe. The final report of the Devonshire Commission (1875) recognized this and declared, 'the progress of scientific research must in a large degree depend upon the aid of the Government. As a nation we ought to take our share of the current scientific work of the World.' Only Edinburgh and London Universities offered degrees in science. Germany was producing four and a half times as many investigators and six times as many scientific papers in chemistry as Britain every year (and similarly in physics, biology and geology). The Commission urged the Treasury to increase its provision for science. This prompted (perhaps better, 'shamed') the government to provide another £4000 a year to the Royal Society. Slowly the principle of providing for institutions was accepted. The Devonshire Commission changed the face of British science in many ways, most notably by establishing the principle of giving direct and stipendiary grants to scientists, as well as increasing the regular Parliamentary Grant to the Royal Society.

There was a gradual acceptance of the propriety and necessity of endowing science, although not all was plain sailing. A Royal Commission on Vivisection linked biological and physiological research with animal experimentation. If biology bred brutality, physics and chemistry brought materialism. This meant that all scientists had to be regarded as suspect – not least when they gave stupid advice. At much the same time that the Devonshire Committee was deliberating, a Royal Commission on Sea Fisheries reported (1866). The ubiquitous Thomas Henry Huxley was a member. He argued that fisherman should be allowed to fish, 'where they like, when they like, and as they like', a *laissez-faire* conclusion that led to the removal of all existing restrictions on sea fishing. It was a view that contradicted the experience of the fishermen themselves, many of whom gave witness to the Commission.

However, a serendipitous event the following year helped to correct this anarchy. In 1867 Anton Dohrn, a German marine biologist, visited an amateur *par excellence* – David Robertson, a Glasgow shop-keeper who had retired to Millport on the small Scottish island of Great Cumbrae, where he spent his time studying the local marine fauna. Here Dohrn realized the enormous advantage of working in an area with abundant supplies of appropriate organisms available for research. He dreamt of establishing 'zoologist stations' in different countries, which would be able to carry out research and also supply specimens to other stations. The British Association appointed a committee to examine this idea; this led to Marine Biological Associations being set up in England and Scotland 'to establish one or more laboratories where accurate researches may be carried on leading to the improvement of zoological and botanical science, and to an increase of our knowledge as regards the food, life, conditions, of British food fishes, and molluscs in particular, and the animal and vegetable resources of the sea in general'. From the start, the government was the largest single contributor to the Associations. Dohrn himself set up the Stazione Zoologica in Naples, which became internationally renowned. Over a brief period towards the end of the nineteenth century, financial provision for research moved from private sources and a small amount of money from learned societies to massive state patronage.

The Royal Society took little part in these developments. Its ethos promoted suspiciousness of 'amateurs' and it was equivocal about supporting them. William Flower, a future Director of the Natural History Museum, declined to serve on a Royal Society committee disbursing grants on the grounds that government money was being dissipated to amateurs at a time when professionals in the nation's museums were labouring under unsatisfactory conditions. Much biological field science was carried out by amateurs, but more and more amateur interest and involvement in science became concentrated in non-mathematical, non-experimental disciplines – such as geography, geology, zoology, botany and meteorology. A consequence of this was that the original meanings of the terms 'naturalist' and 'natural history' were downgraded. A naturalist became a mere descriptive investigator; by the end of the century it had definite associations with amateurism. Arthur Thomson dismissed natural history as a kindergarten subject (ironically in his inaugural lecture as Professor of Natural History in Aberdeen); 'proper' science involved measurement and quantitative rigour. Ernest Rutherford, Nobel laureate for his work on radioactivity, proclaimed that science was either physics or stamp collecting.

This disparaging of natural history took place at the same time as interest in it was burgeoning. In 1873 there were at least 169 local scientific societies in Britain, of which two-thirds were professedly field clubs with a total membership of nearly 50,000. This had doubled to an estimated 100,000 by the end of the century. They represented a vast army of local naturalists with an intimate knowledge of the biology, geology and geography of their localities. The original small societies began to join in regional groupings. The oldest and most successful of these was in Yorkshire, when in 1877 twenty-seven societies (rising to thirty-eight by 1883) joined to form a federal Yorkshire Naturalists' Union, with its own journal – the *Naturalist*. Within each regional union, sectional and research committees were formed, following the pattern which had developed at the national level in the British Association for the Advancement of Science. There were attempts to organize the societies nationally, but these never achieved much impact. The British Association tried to help. A committee chaired by Darwin's cousin, Francis Galton, agreed in 1884 'to draw up suggestions upon methods of more systematic observations and plans of operation for local societies, together with a more uniform mode of publication of their work'. This led to the British Association forming a Conference of Delegates of Corresponding Societies, intended as a national forum for scientific leadership in the societies, with the aim of bringing together professionals with representatives of the societies. An annual conference was held, intended to stimulate local research. In 1888 there were fifty-five Corresponding Societies with 19,000 members. However, less than half the societies used to attend the Conference and, worse, many of the more active (and publishing) local societies did not affiliate. It never really fulfilled its purpose as a network of constituents for the British Association. In truth, amateur field naturalists were neither well equipped nor interested to act as local missionaries for professional science.

The word 'ecology' was first used by Ernst Haeckel in German. At a British Association meeting in 1893 a physiologist, John Burdon-Sanderson (great-uncle of J.B.S. Haldane – p. 16), suggested the need to recognize the content and practice of ecology as a legitimate branch of biology, coequal with morphology and physiology (and, he noted, 'by far the most attractive'). The first major text in plant ecology was *Plantesamfund: Grundtrak af den Okologiske Plantegeografi* by the Professor of Botany in Copenhagen, Eugene Warming, originally published in Danish, but translated into German (1896) and English (1909). This, of course, built on earlier work, perhaps most notably that by Alexander von Humboldt.

In Britain the potential of the army of amateurs was an obvious advantage for initiating a more systematic survey. Arthur Tansley, who was to bestride botany and conservation in Britain for much of the twentieth century, wrote in 1904, 'Scattered up and down the country are scores of men [sic] whose hobby is botany and whose acquaintance with their local floras is absolutely unequalled.'[3] The obvious way to give rigour to all this dispersed and unorganized knowledge was through vegetation mapping. A 'Central Committee for the Survey and Study of the British Vegetation' was formed in late 1904, consisting of nine men, all college lecturers except one – Robert Lloyd Praeger (p. 128). Its first maps were published with the help of grants from the Royal Society and the Royal Geographical Society.

[3] The problems of ecology. *New Phytologist*, 3(8):191–200, 1904.

In 1913, this committee formed the British Ecological Society, with the committee members as its first Council. The Ecological Society of America (ESA) was formed two years after the British one. Like its British equivalent it had its roots in biological survey, but its base was narrower, deriving mainly from agricultural college teachers, although it included zoologists and botanists from the start. The initiative for its formation came from Robert Wolcott, Professor of Zoology in the University of Nebraska, who suggested it to Victor Shelford at the University of Chicago, another zoologist who became the first President of the ESA. From its beginning, the ESA was an assembly of professionals, a high proportion being foresters. Both Societies produced specialist journals. The split between amateur and professional widened.

German scholarship laid stress on priority and the consequent need to authenticate one's findings in print. The custom grew up for the heads of university departments to found their own specialist journals to accommodate their personal output, plus that of their associated staff and pupils. British biology had hitherto appeared almost exclusively in books or society journals, but a spate of new publications began to appear, like the *Journal of Physiology* (1878) and the *Annals of Botany* (1887). Non-professionals, unaccustomed to reading and understanding scientific texts, found it increasingly difficult to keep up. The jargon became ever denser and assumed too much specialized knowledge. For the first time, people found themselves unable to understand text in their native tongue. The 'Two Cultures' were born.

BOX 6.4   **The Two Cultures**

History is full of accounts of clashes between cultures – some involving strife, many persisting for long periods, although most temporary and trivial. The dispute between Galileo and the ecclesiastical establishment of his time was an early disagreement about the degree of authority that could – or should – be accorded to observational data. A later and repeatedly misreported one was the

FIGURE 6.2 Assumed attitudes. Thomas Huxley and Bishop Samuel Wilberforce debated together in 1860 in Oxford. Their debate is routinely misreported as a triumph for Huxley and a humiliation for the bishop. In fact the two protagonists had different concerns and argued across rather than against each other. It is said that the audience regarded it as a draw.
Cartoons from *Vanity Fair.* Photo: Oxford Science Archive/Print Collector/Getty Images.

verbal fisticuffs between the Bishop of Oxford and Thomas Huxley at the Oxford meeting of the British Association for the Advancement of Science in 1860, with the Bishop seen to be behaving Canute-like in the face of scientific advance and the heroic Huxley triumphing over the forces of reaction (Figure 6.2).

The truth was very different. The Bishop was certainly pompous, but he was no blinkered fool. He had a first-class degree in mathematics and was a Vice-President of the British Association. His concern was about the dangers of change:

- sociological – traditional church-going was being disrupted by the movement of people into towns, where they were less likely to be worshippers;

- theological – the apparently unstoppable successes of manufacture, engineering and colonialism contrasted and seemed to conflict with the biblical picture of human weakness and the need for redemption; and
- ecclesiological – questionings of the authority of the Bible (and hence, the Church) were beginning to spread, fuelled by the 'higher criticism' of German scholars and fed by the recent publication in England of *Essays and Reviews*, a collection of essays by authors who rejected the verbal inspiration of scripture.

Bishop Wilberforce's attack on evolution was secondary to his main aim of protecting the status quo of society and Church.

Likewise, Huxley's main target was not religion or Christianity but what he regarded as the illegitimate authority of Church leaders, holding on to offices of secular power and thus excluding the voice of science, which to Huxley represented objective truth as opposed to superstition. His aim was (and remained throughout his life) the secularization of society. The disputants were not so much arguing against each other as across each other.

There are no verbatim accounts of the meeting. Huxley was certain he had 'won' and that he was 'the most popular man in Oxford for four and twenty hours afterward'. In contrast, Joseph Hooker (who spoke after Huxley) reported to Darwin that Huxley had spoken too fast and was largely inaudible, and that it was he (Hooker) who had carried the day. The Bishop went away happy that he had given Huxley a bloody nose. As far as the audience were concerned, many scored it as an entertaining draw. Notwithstanding, the debate is repeatedly recalled as showing the inevitable impotence and irrelevance of religion faced with 'true' science.

The debate and its accompanying misconceptions show no sign of dying. The 1860 Oxford debate is perhaps the most notorious, but scarcely less famous is the 1926 Scopes trial in Dayton, Tennessee, when a young school teacher was prosecuted for teaching evolution, which was against State law. Once again the battle was really about non-science issues – the self-promotion of a small town, the freedom of expression under the American Constitution, the agenda of a populist politician (William Jennings Bryan) and the ambition of the American Civil Liberties Union for a test case. Helped by a Broadway

FIGURE 6.3 Manufactured attitudes. In 1925, John Scopes, a rather naïve school teacher, was tried and found guilty in the State of Tennessee for teaching human evolution, which was illegal there. The trial highlighted very different (and still persisting) attitudes towards science and religion in the USA.
Photo: Topical Press Agency/Stringer.

play later filmed as *Inherit the Wind*, it is another formidable myth celebrating the triumph of science over prejudice (Figure 6.3).[1]

'The Two Cultures and the Scientific Revolution' was the title of a lecture given by the physical scientist and author C.P. Snow in Cambridge in May 1959. He described his experience:

> A good many times I have been present at gatherings of people who, by the standards of the traditional culture, are thought highly educated and who have with considerable gusto been expressing their incredulity at the illiteracy of scientists. Once or twice I have been provoked and have asked the company how many of them could describe the Second Law of Thermodynamics. The response was cold: it was also negative. Yet I was asking something which is the scientific equivalent of 'Have you read a work of Shakespeare's?'... Constantly I felt I was moving among two

groups – comparable in intelligence, identical in race, not grossly different in social origin, earning about the same incomes, who had almost ceased to communicate at all, who in intellectual, moral and psychological climate had so little in common that instead of going from Burlington House [where the Royal Society was based] or South Kensington [home to the Science and Natural History Museums and Imperial College] to Chelsea [and its bohemian intelligensia], one might have crossed an ocean. In fact, one had crossed much more than an ocean – because after a few thousand Atlantic miles, one found Greenwich Village talking precisely the same language as Chelsea, and both having about as much communication with MIT as though the scientists spoke nothing but Tibetan.

Snow's thesis was condemned by some – most violently by the literary critic F.R. Leavis – intriguingly repeating a similar dispute almost a century earlier between Thomas Arnold [sometimes seen as a bridge between Romanticism and Modernism] and his relative by marriage, Thomas Huxley. The debates still reverberate.

[1] Larson, E.J. (1997). *Summer for the Gods*. Cambridge, MA: Harvard University Press.

Meanwhile Huxley dominated the development of biology and biological teaching in Britain. He devised the practice of concentrating on 'types' instead of a full context of the lives of animals and plants. Huxley himself was clear about his limitations, noting, as we have seen, in his *Autobiography* (1899), 'There is very little of the genuine naturalist in me.' This meant that generations of biologists were raised in an environment where the emphasis was almost entirely on a very limited understanding of structure and function. They became not simply biologists, but narrowly focused specialists and militantly anti-amateur. The efforts of Huxley and his X Club friends had the effect of fragmenting specialists and their associated societies in a way that did not previously exist. Unlike chemistry and physics, which largely retained their unitary structure, biology fragmented into an explosion of reductionist disciplines, a trend which has become even more acute with the excitements and allure of molecular science.

FURTHER READING

Allen, D.E. (1998). On parallel lines: natural history and biology from the late Victorian period. *Archives of Natural History*, **25**: 361–71.

Attenborough, D. (2002). *Life on Air*. London: BBC Books.

Hewitt, R. (2010). *Map of a Nation: a Biography of the Ordnance Survey*. London: Granta.

Lowe, P.D. (1976). Amateurs and professionals: the institutional emergence of British plant ecology. *Journal of the Society for the Bibliography of Natural History*, **7**: 517–35.

MacLeod, R.M. (1971). The support of Victorian science. The endowment of research movement in Great Britain, 1968–1900. *Minerva*, **4**: 197–230.

Meadows, J. (2004). *The Victorian Scientist: the Growth of a Profession*. London: British Library.

Parsons, C. (1982). *True to Nature*. Cambridge: Patrick Stephens.

Sands, T. (2012). *Wildlife in Trust: a Hundred Years of Nature Conservation*. London: Elliott & Thompson.

Sobel, D. (1996). *Longitude*. London: Fourth Estate.

7   National Nature – a Digression

Britain has had laws to protect game and other wildlife for as long as it has had a parliament to make them. The forest law of the Norman kings regulating hunting and land use in the 'Royal Forests' was a major feature in the development of landscape over much of England. About 200 years ago the pace of legislation quickened when agricultural reform and increases in human numbers produced pressures on land. Enclosing tracts of land was a necessary preliminary to improving farmland (see Chapter 3, pp. 43–62). Enclosure originally took place through informal agreement, but the practice developed in the seventeenth century of needing authorization through an Act of Parliament. Over 700 such acts were passed between 1800 and 1850. During the same period there was a growing movement of people into towns; any available land near an industrial town was likely to be a tempting building site. A rather belated safeguard was introduced in 1845 with a General Enclosure Act, which accepted the idea that enclosure was the concern of all the local inhabitants, not merely a privileged few; it decreed that the health, convenience, exercise and recreation of everyone should be taken into account before any enclosure was sanctioned. In 1865, the Commons, Open Spaces and Footpaths Society was formed to resist the continuing enclosure of common land. Its successes included saving Hampstead Heath, Epping Forest, Wimbledon Common, Ashdown Forest and the Malvern Hills. The society was the forerunner of a host of pressure groups to protect or conserve the environment.

One of the solicitors for the Commons Society was Robert Hunter. He made common cause with Octavia Hill, a formidable social reformer who had been inspired by John Ruskin and the Christian socialist F.D. Maurice about the lack of fresh air and green space in London. They were joined by an impatient young Church of England clergyman, Hardwicke Rawnsley, to campaign against the planned damming of the central Lake District lake Thirlmere by Manchester Corporation for use as a reservoir. The three of them went on to be the driving force for the establishment in 1893 of a National Trust for Historic Sites and Natural Scenery (now simply the 'National Trust'), originally envisaged as a land company to buy and accept gifts of land, buildings and common rights for the benefit of the nation, without judgement as to their wider importance. It had a natural ally in the Society for the Prevention of Cruelty to Animals (later the Royal Society; RSPCA), founded in 1824 to campaign against cruelty to domestic animals such as cows and horses, but later including bear-baiting and cock-fighting. Concern about the wholesale slaughter of great crested grebes and kittiwakes to provide plumes for women's hats led to a Sea Birds Preservation Act of 1869 and a Wild Birds Protection Act of 1880, followed in 1889 by a Society for the Protection of Birds (now the RSPB), set up by a group of formidable women, led by the Duchess of Portland, upset by the plight of young birds left to starve after their parents were shot for plumes. Then in 1885 the more inclusive Selborne Society for the Protection of Birds, Plants and Pleasant Places was formed, usually regarded as the first nature protection body.

Most of these efforts were concerned with protecting individual species. The next stage was looking after their habitats. The initiative for this came from two very different people: Charles Rothschild (1877–1923), a scion of the Rothschild banking family, and Frank Oliver, Professor of Botany at University College London. In 1899, Rothschild purchased Wicken Fen, a relic of the once widespread fens of East Anglia, and donated it to the National Trust, with the stipulation that it be managed as a nature reserve; he was convinced of the need to manage 'good spots' as nature reserves (Figure 7.1). Wicken Fen has to be actively managed by controlling the water level which is much lower in surrounding agricultural areas. Then in 1912 he founded the Society for the Promotion of Nature Reserves (SPNR) to

FIGURE 7.1 Wicken Fen, a remnant of a formerly widespread habitat, now a national nature reserve. It supports many flowering plants and a rich invertebrate fauna, which are now uncommon elsewhere.
Photo © Andrew Stawarz.

identify and preserve 'areas of land which retain primitive [i.e. ancient and semi-natural] conditions and contain rare and local species liable to extinction owing to building, drainage, disafforestation, or in consequence of the cupidity of collectors'. In the same period Oliver, together with his junior colleague Arthur Tansley, began taking student parties to study littoral habitats, at first in Brittany and then at Blakeney Point in Norfolk – which he bought in 1912, helped by a substantial gift from Rothschild.

The following year Oliver, recalling the establishment of the SPNR, challenged ecologists to consider the establishment and potential value of nature reserves, pointing out that the few nature reserves in England formally recognized at the time existed more by accident than any deliberate policy, perhaps because 'the country districts of England are not obviously and seriously threatened, hence the Nature Reserve movement lacks any background for a strong public appeal'.[1] Rothschild was less sanguine about this lack of

[1] Nature reserves. *Journal of Ecology*, **2**, 55–6, 1914.

danger. Under his leadership, the SPNR consulted widely, producing a list of 273 areas 'worthy of protection'. It submitted this list to the government's Board of Agriculture in 1915. But there was little interest and the SPNR itself became less active after the premature death of Rothschild in 1923. The St Kilda archipelago (now a double World Heritage Site) was offered to the Society for £3000 in 1927, but 'after some discussion the Committee decided to take no action as the scheme was so large and it did not appear that the fauna there was in serious danger'. Most debate in the interwar years centred on the establishment of large national parks, which were supposed to include nature conservation in their remit, although doubts were repeatedly voiced that this would achieve adequate protection.

In 1929 the Councils for the Preservation of Rural England, Wales and Scotland petitioned the Prime Minister (Ramsay MacDonald) for an enquiry about the need for national parks. The Prime Minister responded by appointing a committee under physician and politician [Lord] Christopher Addison to advise him. The committee recommended that such national parks should indeed be established, to:

- safeguard areas of exceptional national interest against disorderly development and spoliation;
- improve the means of access to areas of natural beauty; and
- promote measures for the protection of flora and fauna.

An unlikely step towards fulfilling these recommendations came in 1940 from a Royal Commission on the Distribution of the Industrial Population, which urged the establishment of a central planning board to regulate urban developments. Its publication in early 1940 sparked an awareness of the need to plan for reconstruction after the Second World War ended. This prompted the SPNR to focus on how to do this. In 1941 it convened a Conference on Nature Preservation in Post-War Reconstruction. This gathering argued for the need to incorporate nature reserves into any national planning scheme. The government accepted the concept, but prevailed upon the Conference members to develop a detailed scheme of what was needed. A Nature Reserves Investigation Committee was set under a senior judge, Lord Justice Scott, to assess the impact of this on the well-being of rural communities. His report eulogized the English landscape as 'a striking example of the interdependence between the satisfaction of man's material wants and the creation of beauty'. The British Ecological Society supported the Scott Committee with a rationale for the existence of nature reserves and compiled a list of forty-nine possible national habitat reserves and thirty-three scheduled areas, plus eight sites where wildlife was already protected, 'places suitable for preservation, with relevant information about them'. The Society emphasized that the most important aim of the preservation movement was 'the maintenance for enjoyment of the people at large of the beauty and interest of characteristic British scenery… (touching) the deepest source of mental and spiritual refreshment, both conscious and unconscious'; the scientific, educational and economic values of preservation were essential but secondary to this.[2] The Scott Report (1942) formed the basis for rational land use and planning, formalized in a Town and Country Planning Act (1947) and a National Parks and Countryside Act (1949). The latter was presented to Parliament as enabling access to the countryside, which is 'just as much a part of positive health and wellbeing as are the buildings of hospitals or insurance against sickness'. It was largely the fruit of yet another Committee – on National Parks, under Arnold Hobhouse. This in turn spawned two Wild Life Conservation Special Committees, one chaired by T.H. Huxley's grandson Julian for England and Wales and the other by James Ritchie for Scotland.

---

[2] A policy document issued by the statutory Nature Conservation Council (NCC) in 1984 (*Nature Conservation in Great Britain*) used rather similar language: 'the proper role for the NCC and most of the NGOs [non-governmental organizations] is to practise nature conservation according to a definition of purpose which is primarily cultural, that is the conservation of wild flora and fauna, geological and physiographic features of Britain for their scientific, educational, recreational, aesthetic and inspirational value. The term cultural should not be misconstrued: it is used here in the broadest sense as referring to the whole mental life of a nation.'

These last two committees argued that reserves should be chosen primarily for their scientific value, to include 'both the unique and the typical, the common and the rare, in such proportions as will best provide a foundation for a sound ecological study of wildlife conditions in this country'. The government accepted this and set up a statutory Nature Conservancy, which received a Royal Charter in 1949. Its role was to provide expert advice on nature conservation, to designate and manage a series of national nature reserves and to undertake such research as was relevant to those functions, over and above the more fundamental, long-term research expected of a research council. An Institute of Terrestrial Ecology was split from the Nature Conservancy in 1973 and became part of a newly formed Natural Environment Research Council; the Conservancy, retaining its conservation functions and some research responsibilities, changed its name to Nature Conservancy Council.[3] There is no doubt that the existence of the Nature Conservancy had a synergistic effect on the growth of interest in ecology in Britain.

The Nature Conservancy was very much the province of professionals. As far as amateurs were concerned, a Central Co-ordinating Committee for the Protection of Nature (in part modelled on the Yorkshire Union) had been formed from ten societies in 1924. It gave evidence to the Addison Committee, but in general found little to do and wound itself up in 1936.

In the 1940s and 1950s the SPNR was mainly taken up with the Nature Reserves Investigation Committee and its outcomes, especially the recommendation that there should be a national system of nature reserves. But activists were beginning to focus on their local areas. A Naturalists' Trust had been formed in 1926 to acquire and preserve the important coastal bird site of Cley Marshes on the north coast of Norfolk, near Cromer, which had been offered to but

---

[3] In 1991 the Nature Conservancy Council itself was divided into separate agencies for England, Scotland and Wales, with a Joint Nature Coordinating Committee supposed to link them and deal with international issues.

turned down by the National Trust. By 1941, the Norfolk Trust had acquired eight reserves. In 1938, a Pembrokeshire Bird Protection Society was established, partly initiated by Ronald Lockley who, inspired by a visit he had made to the pioneering work on the North island of Heligoland, had set up the first British bird observatory on the island of Skokholm where he lived. The Bird Protection Society became the West Wales Field Society, which in 1946 bought a coastal fort (built for the protection of a nearby naval dockyard in the 1850s) and leased it as one of the first centres of the Field Studies Council to run as a field centre. Then in 1945 the venerable Yorkshire Philosophical Society acquired Askham Bog south of York, a survivor of the ancient Yorkshire fens, and formed the Yorkshire Naturalists' Trust. These local trusts were followed by others, joining together in 1957 to form a Council for Nature, an association for mutual support and encouragement under the aegis of the SPNR, modelled on the (US) National Audubon Society. Despite misgivings about the danger of submerging local interests, membership of the Council for Nature was opened to all interested local or national natural history societies. Within five years, the Council for Nature had obtained the support of nearly 400 organizations with a total membership of around 100,000.

The formation of the Nature Conservancy freed the SPNR (which became the Royal Society for Nature Conservation in 1981) from its role as a pressure group, allowing it to concentrate on coordinating and stimulating local action. The whole of Britain is now covered by forty-seven Wildlife Trusts, each with their own full-time staff.

The Wildlife Trusts attracted activists. But the wider public were increasingly involved as television became widely available. 'Nature' programmes became popular. These began appearing in the early 1940s before many people had television sets. Early programmes consisted chiefly of animals in zoos. Peter Scott fronted the first BBC wildlife programme in 1953, largely based on film footage he had collected on his own expeditions. This became a regular fortnightly event, presented by Scott for seventeen years. Around the same time, the Disney Corporation started filming wildlife subjects, albeit with an emphasis on entertainment rather than scientific accuracy. Armand and Michaela Denis produced films for a number of TV companies from 1954. In 1957 the BBC set up its own Natural History Unit, which consistently pioneered techniques and films, most famously David Attenborough's nine *Life* series. Some of these programmes attracted audiences of several millions and undoubtedly increased public awareness of the environment, complemented by straightforward documentaries on subjects such as pollution, climate change, extinctions, agricultural practices and so on. RSPB membership swelled to over a million, that of the National Trust to over four million. English Heritage was formed as a charity from a number of governmental bodies charged with looking after historic sites. Nearly a million and a half people are now members of English Heritage and its sister organizations in other parts of the UK.

Another significant event in 1945 was the launch by the publisher William (Billy) Collins, excited by the opportunity of affordable colour printing, of New Naturalist books with 'the aim to interest the general reader by recapturing the enquiring spirit of the old naturalists… fostered by maintaining a high standard of accuracy combined with clarity of exposition in presenting the results of modern scientific research' – in other words, to bridge the gap between amateurs and professionals. A committee under Julian Huxley planned the series, with James Fisher – whose book *Watching Birds* had been a runaway success when it appeared in 1940 – as secretary. The first New Naturalist volume was published in 1945; the series is still going, having reached no. 136 by 2017. The New Naturalist series was the forerunner of many well-illustrated guidebooks, perhaps best typified by the *Field Guide to Birds*, illustrated by Roger Peterson, first published in 1953 and expanded ever since for many groups.

In 1963, the Council for Nature organized a National Nature Week to publicize the increasing threat to wildlife and the part that the various natural history groups could play in countering this. The

FIGURE 7.2 Two pictorial stamps – one celebrating the plant world, the other the animal world – were issued to mark National Nature Week in 1963, the first commemorative stamps to have any sort of environmental theme (left). In recent years commemorative series with an environmental theme have appeared at ever-decreasing intervals. They have included: British birds in 1966, wild flowers in 1967, centenary of Wild Bird Protection Act in 1980, river fish in 1983, centenary of the RSPB in 1989, centenary of RSPCA and 150th anniversary of Kew Gardens in 1990, protection of the environment in 1992, the weather in 2001, insects in 2008, World Wildlife Fund in 2011, butterflies in 2013. Four stamps were issued in 1986 to mark endangered species (right).
Reproduced by permission of the Royal Mail; stamps designed by K. Lily, M. Goaman and S. Scott.

British Post Office issued commemorative stamps, and a 'wild life exhibition' was sponsored by the *Observer* newspaper (Figure 7.2).

One of the visitors to the exhibition was the Duke of Edinburgh, who was impressed with the way the event brought together groups who normally met rarely. It was a time of rising agricultural productivity, encouraged by the need to pull agriculture out of recession and reduce rural poverty – but which had the effect of changing the countryside in many ways through the use of chemicals and by rooting out hedges and creating ever-larger fields. The Duke led an effort to build on the success of the exhibition through a study conference, to be called the 'Countryside in 1970'. In fact, three such conferences were held – in 1963, 1965 and 1970. The Duke was clear about the aims of this initiative: that conservation was not a way of putting the clock back; there was no intention of turning the countryside into an open-air natural history museum. The endeavour marked a major step forward in the practice of nature conservation. In the words of the secretary of the conferences, Max Nicholson, it meant 'the replacement of narrow concepts of protection of animals and plants through laws and sentiment and of natural history and love of nature as merely hobby interests, by a recognition of man's care for the natural environment as an essential concern of government and of civilised society generally'. Nicholson saw the conferences as marking the beginning of an 'environmental revolution' involving:

> a new sense that a 'good' or 'high-quality' or 'healthy' environment is not only more pleasant but may be essential to a sound and viable civilisation. Closely linked with that sense is a new awareness that we now have it in our power to conserve or to damage and destroy the countryside on a vast scale... Part of the significance of the 'Countryside in 1970' conferences comes from the making of a loose triple alliance between the nature conservationists, the amenity and rural preservation movement, and the outdoor recreational interests which have almost simultaneously felt the supply of their fundamental requirements threatened by explosive processes of development and of population expansion.[4]

The government recognized the importance of the conferences by widening the remit of the existing National Parks Commission into a statutory Countryside Commission (now merged with the devolved conservation bodies – Natural England, Countryside Council for Wales and Scottish Natural Heritage); the voluntary bodies formed a Council for Nature to focus attention on the competing interests in the countryside and the need to reconcile these.

[4] *The Quarterly Review*, **304**: 121–30, 1966.

BOX 7.1   **Max Nicholson**

Nicholson (1904–2003) was an immensely important figure in shaping environmental attitudes in the twentieth century, both in his native Britain and internationally as one of the founders of both the International Union for the Conservation of Nature and the World Wide Fund for Nature (Figure 7.3). His interest in the natural world began at the age of seven with a visit to the Natural History Museum in London, and developed through reading the *Natural History of Selborne* and books on the English countryside, such as those by W.H. Hudson (1841–1922) and Richard Jefferies (1848–87). As a student reading history at Oxford University, he organized bird counts and was the driving force behind the formation in 1933 of the British Trust for Ornithology to coordinate surveys of birds by drawing on observations by both amateurs and professionals. He

FIGURE 7.3 Max Nicholson (1904–2003), a driving force in environmental conservation and shaping environmental attitudes throughout most of the twentieth century.
Photo: Roy Jones/Evening Standard/Hulton Archive/Getty Images.

claimed he learnt as much from bird-watching as from everything else he was taught at university. He published his first book (*Birds in England*) in 1926, the year he entered university. As a civil servant after completing his university degree, he was largely responsible for drafting the legislation which led to the establishment of the statutory British Nature Conservancy. He was Director-General of the Conservancy for fifteen years (1952–66) and in this role played a major part in defining and developing the idea of science-based nature conservation. He described his approach to the natural world:

> Unsullied nature is majestic, fascinating and varied, but also uncompromisingly impersonal, mindless and somewhat chilling... It is obvious that nature can be impoverished and mutilated by man, and equally that through talented landscaping and careful management it can be transformed into something more congenial and attractive to many than its primitive state... Yet to a purist both these types of treatment are in principle equally unpalatable, in that they replace the spontaneous natural community with a transformed substitute, or an artefact.[1]

[1] *The Environmental Revolution* (1979). London: Hodder & Stoughton, p. 20.

Max Nicholson was effectively the driving force behind the Council for Nature, which he envisaged as an umbrella body to bring together field natural history and conservation interests. It initially functioned well, but it faced internal tensions – between the Council and the growing strength of the regional (county) Naturalists' Trusts which had linked into the SPNR as a national body, and between the Council and the increasingly powerful RSPB. In 1979 these tensions, compounded with a shortage of finance, led to the Council for Nature disbanding. Some of its functions were transferred to the Council for Environmental Conservation (CoEnCo). The coordinating functions were passed on to the newly formed Wildlife Link in 1980 – a looser coalition of more politically active conservation bodies. Ten years later Wildlife Link joined a sister group, Countryside Link to form Wildlife and Countryside Link.

Despite the lack of cohesion, concern and action for environmental protection continued to increase. Wildlife Link and the official Nature Conservancy under Max Nicholson exerted pressure on the British Government to enact a much stronger Wildlife and Countryside Act than was first envisaged, one which joined habitat and wider countryside issues and established a more formal system for establishing nature reserves. The legislation was shaped by British pressure but also by the need to respond to the European Birds Directive adopted by the European Union in 1979 (which was followed and supplemented by a Habitats Directive two years later, which established objectives for nature for which the British Government would be held accountable).

Meanwhile, increasing awareness of damage from agricultural chemicals chronicled by Rachel Carson in her best-selling book *Silent Spring*, published in 1962, achieved a rare political alignment in the USA, getting support from Republicans and Democrats, rich and poor, city slickers and farmers, tycoons and labour leaders (p. 187). This swell of support contributed to the first 'Earth Day', celebrated in the USA on 22 April 1970, as the response of Wisconsin Senator Gaylord Nelson to a massive oil spill off Santa Barbara, California. The enthusiasm generated by Earth Day led to the creation of the US Environmental Protection Agency and the passage of the Clean Air, Clean Water and Endangered Species Acts. In Europe, impressed with the National Nature Week in Britain, the Council of Europe declared 1970 as European Conservation Year. Friends of the Earth and Greenpeace were both established in 1971.

These national initiatives gave impetus to international efforts, which had been gathering pace since the late nineteenth century. US President Theodore Roosevelt, encouraged by conversations with John Muir, was posed to convene an International Conservation Conference in 1909, but his successor, W.H. Taft, cancelled this and also dismissed the conservationist Pinchot as head of the Forest Service. The Sierra Club, founded in May 1892 in San Francisco by John Muir, who became its first president, lost steam

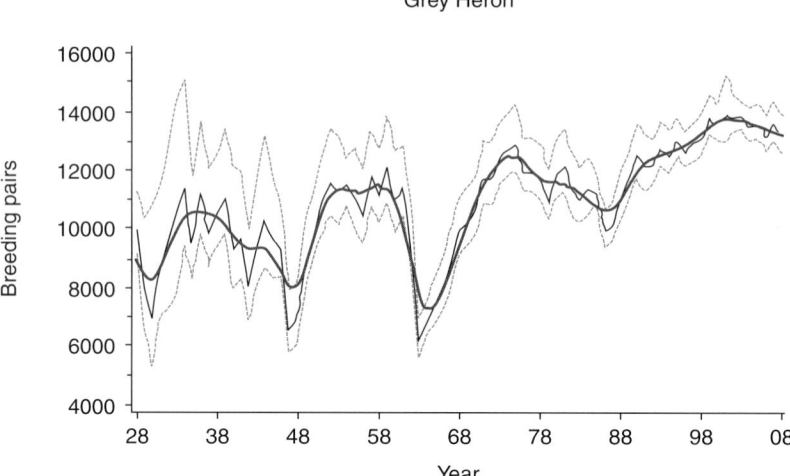

FIGURE 7.4 Fluctuations in numbers of herons (*Ardea cinerea*) in the UK, the longest running census of any bird species.
Reproduced with permission from Robinson, R.A., Leech, D.I., Massimino, D. et al. (2016). *BirdTrends 2016: Trends in Numbers, Breeding Success and Survival for UK Breeding Birds. Research Report 691.* Thetford: British Trust for Ornithology.

after Muir died in 1914. Effective US leadership in the area lapsed for several decades. However, international linkages were beginning to form. An International Committee for Bird Protection (now BirdLife International) was established in London in 1922. It was the product of concern about the trade in wild bird feathers and the threats to migratory birds across several continents as seen by the RSPB (founded 1889), the Audubon Society of the USA (founded 1905), Jean Delacour from France, and P.G. van Tienhoven and A. Burdet from Holland. Their vision was that united action was likely to be much more effective than national organizations working individually. Birds are obvious – almost emblematic – indicators of the state of the natural world (Figure 7.4). Deaths of sea birds from oil was publicized by the RSPB in Britain and paralleled by concern in the USA. It led to an attempt in 1926 to legislate about the discharge of oil at sea. It failed because of objections from Germany, Italy and Japan.

BOX 7.2 **Ornithological publishing**

The first book devoted to (mainly) British birds appeared in 1544. It was written by William Turner (p. 66), a friend of Konrad Gesner and better known as a botanist. It was called *A Short and Succinct History of the Principal Birds Noticed by Pliny and Aristotle*. It listed about 120 species, most of them occurring in Britain, despite the title. However, the first comprehensive list of British birds did not emerge for another century, soon to be followed by Willughby and Ray's *Ornithologia libri tres* (1676). It took another hundred years for the next major work, Thomas Pennant's *British Zoology*, to be published (1766–7). It was based on Pennant's own observations, supplemented by his correspondents and enhanced by the inclusion of coloured plates. These plates were such a successful feature of another of his literary specialties, travel writing, that it has been suggested that they were instrumental in the development of tourism (Figure 7.5). Gilbert White corresponded with Pennant but

FIGURE 7.5 Dunvegan Castle on Skye, from Pennant, T. (1774) *A Tour in Scotland and Voyage to the Hebrides*, a book which effectively gave birth to mass tourism.
Artist: Peter Mazell. Photo: The Print Collector/Getty Images.

contributed more importantly by the enquiring approach he practised and communicated, rather than the mere 'identify and describe' habits of his predecessors and contemporaries.

The ornithological world was infected by movements elsewhere. Hunters liked to preserve their trophies and often asked local tanners to cure the skins and stuff them, a practice which developed into taxidermy. Collectors began assembling specimens; museums acquired ever more representative collections. The study of such collections became the province of professional taxonomists, who were divorced from those who studied birds in the field and who speculated about the mysteries of migration and navigation.

More inclusive action was necessary. Birds represent only a small segment of the natural world. External events conspired with internal realizations to produce a cascade of appreciation of needs over a wide spectrum. This emerged in 1948 with the setting up of an International Union for the Protection of Nature (now the International Union for the Conservation of Nature and Natural Resources; IUCN) through the actions and advocacy of the French Government, the Swiss League for Nature Protection and, crucially, the United Nations Educational, Scientific and Cultural Organization (UNESCO), led by Julian Huxley, its first Director-General, who had insisted that 'culture' without 'science' was meaningless. The scope of the IUCN was broad, involving both governments and non-governmental agencies. By the end of the twentieth century, awareness of environmental problems in at least western Europe and North America was greater than it had been since the pioneering days of modern agriculture in the sixteenth century.

FURTHER READING

Holdgate, M. (1999). *The Green Web*. London: Earthscan.

Parsons, C. (1982). *True to Nature*. Yeovil: Patrick Stephens.

Sands, T. (2012). *Wildlife in Trust*. Newark: The Wildlife Trusts.

Sheail, J. (1998). *Conservation in Britain: the Formative Years*. London: Stationery Office.

Smith, T. (2007). *Trustees for Nature*. Horncastle: Lincolnshire Wildlife Trust.

8 The Regulatory Century

The second half of the twentieth century led to ever more structuring of the knowledge and regulation of the environment. It produced a packaging but also a sanitation of environmental attitudes. On the one hand, there was an increasing awareness of threats to the natural world from agencies in the environment itself – running out of oil, climate change, species loss, pollution damage – but on the other hand, it all seemed to become someone else's problem. In the developed world, the average person (assuming there is such a person) was insulated from his or her surroundings in heated (and often air-conditioned) houses, able to travel in increasingly reliable cars or comfortable planes, protected against previously dangerous diseases and even from 'acts of God' by modern technological know-how. Actually, this protective cocoon applied to only a minority of people living in only a few favoured countries, but they were the people with the communication skills to persuade themselves that this was the possible and acceptable condition for all. 'Nature' was experienced second-hand through television or ever-easier tourism. Uncertainties in the real world were taken care of by external, often international agencies, topped up by insurance so that if something did go wrong, there was generous financial recompense.

BOX 8.1 **Acts of God**

The number of people affected by disasters (i.e. requiring basic survival needs such as food, water, shelter, sanitation, immediate medical assistance) is rising steadily. Over 200 million people are affected every year, despite technological advances in building and identifying hazardous areas from flooding or earthquakes. 'Natural disasters' are best defined as the result of unexpected events, rather than expected but rare ones. Earthquakes are caused when the tectonic plates which make up the Earth's crust move against each other, producing stresses in the rocks. An earthquake occurs when these tensions are released, producing energy which ripples out as seismic waves. They arise most often around the boundaries of tectonic plates. In themselves they rarely affect human life. Their dangers are virtually all due to unpreparedness or secondary effects. Perhaps the most notorious earthquake of all occurred in Lisbon on All Saint's Day 1755, and killed over 40,000 people. Most of the deaths were due not to the earthquake itself, but to the fire which followed the actual earthquake, lit by candles in the church services taking place when the earthquake struck, and rapidly spreading through the tightly packed wooden buildings. A massive earthquake in Japan in 2011 killed around 19,000 people, despite the fact that earthquakes are common in Japan and buildings are designed to resist them. In this case, almost all the deaths came from a tsunami which followed the earthquake itself.

There have always been 'natural' disasters with massive impacts (p. 56). An underwater landslip off the Norwegian coast around 6000 years ago caused a massive tsunami which extended 80 km inland in eastern Scotland and played a part in submerging the remnants of the land between Britain and continental Europe. What had been a peninsula of Europe became the island of Great Britain with its own constellation of lesser islands. Enemies 'within' could be just as devastating. The Chinese have suffered 2000 famines in the past 2000 years; an estimated five to ten million died from starvation in the USSR in 1918–22 and 1932–4; four million in China in 1920–1; two to four million in West Bengal in 1943. Even in Britain, there were 200 famines between AD 10 and 1846. Deep ploughing and the failure to husband moisture on the friable soils of the USA and Canadian prairies during droughts in the mid-1930s led to massive wind erosion and loss of soil over a quarter of a million hectares, a so-called 'Dust Bowl'. Three and a half million people had to leave the affected areas.

Time after time cultures have contributed to their own decline. The early Polynesian population of New Zealand depended on large flightless moas for food, but after 600 years they had cleared so much forest that a number of bird species, including swans, eagles and moas, were virtually extinct. At some stage there must have been an 'ecological crisis' when the increasing difficulty of obtaining enough moas implied inadequate recruitment of young birds into the population as the number of adults available to be hunted decreased.

The decline of the great Babylonian grain-growing civilization was probably due to the increasing salinity of irrigated areas as a result of imperfect drainage. Again, there must have been a time when yields were declining, more and more unsuitable areas were being pressed into cultivation and irrigation channels required extending and reconstruction, while at the same time the demands of the cities for food would have been increasing.

Easter Island, a small piece of land (64 square miles or 17,000 ha) in the central Pacific, is often quoted as a classical example of overexploitation and misuse of finite resources. The island was first settled by humans around 1000 years BC, with a population which had grown to about 15,000 by 1600. It was well wooded, with twenty-one species of trees identified from pollen grains. However, the island was largely denuded of trees and all the terrestrial bird species killed, apparently by a combination of harvesting for firewood and boat-building, plus the ravages of the Polynesian rat (*Rattus exulans*), intentionally introduced on many islands (probably including Easter Island) for food. The lack of wood for boats made fishing impossible. Removal of trees led to soil erosion. There seems to have been fighting between settlements. When the first Europeans arrived in 1722, the population had plummeted to around 3000. This account of decline through linked causes has been disputed by some, but seems largely correct and a cautionary tale of living beyond one's means.

There are many other examples. Sicily was once the 'granary of Italy', but less and less corn grew there as the soil deteriorated under excessive cultivation and browsing by goats. The usual responses to crises are to overcome them either by emigration (as from Ireland after the years of potato blight in the 1840s) or through the introduction of new technologies (such as the Haber Process for producing nitrogen fertilizer from 1913 onwards, or the 'green revolution' from high-yielding strains of grain in the 1960s). While it is true that saving innovations have usually appeared in the past – ranging from domestication and the Neolithic Revolution to the use of fossil fuels ('pickled sunlight') in modern societies, it is obviously dangerous and disingenuous to rely on novel developments emerging when and where needed. Many communities – and, indeed, civilizations – have disappeared when effective solutions have not emerged. The only other alternative is to plan, organize and regulate – and this involves humility and cooperation. Indeed, if *Homo sapiens* is going to live up to its name as a 'wise animal', regulation seems a proper complement to the serendipitous appearance of useful inventions.

Is 'taking thought for the morrow' on this scale practicable or unnecessary hysteria? After all, we are a very recent incomer to the Earth. Notwithstanding, we are certainly making a major impact on it. Evidence of human influence is everywhere: land-use patterns are visible from space, and concentrations of carbon dioxide, methane, nitrous oxide and other gases are increasing in the atmosphere as a consequence of human activities. It is sometimes suggested that we have moved from the Holocene epoch to the Anthropocene, in which human activities have begun to have a significant global impact. Some say this began in the first millennium after Christ with the modification of soils. Others date it from 1610, with the European colonization of South America and the spread of disease, intercontinental trade, etc. The erosion of species boundaries by human-mediated dispersal of species into new regions and their subsequent naturalization has certainly been a major sign of human impact. Still others want it to start from 1945. The date does not matter. Our effect is indubitable. The important question is whether we ought to have a role in reducing it, re-engineering the keystone, as it were. Harvard palaeontologist Stephen Jay Gould regarded it as arrogant to think that we, the human species, can affect the natural world in any significant way. He wrote about the views:

> that we live on a fragile planet now subject to permanent derailment and disruption by human intervention [and] that humans must learn to act as stewards for this threatened world... however well intentioned, are rooted in the old sin of pride and exaggerated self-importance. We are one among millions of species, stewards of nothing. By what argument could we, arising just a geological microsecond ago, become responsible for the affairs of a world 4.5 billion years old, teeming with life that has been evolving and diversifying from at least three-quarters of this immense span? Nature does not exist for us, had no idea we were coming, and doesn't give a damn about us.[1]

A more radical approach is the Gaian notion that all the Earth's processes are so tightly linked together that they form a self-regulating equilibrium, so that any disturbing human influences are buffered and therefore unimportant and ignorable. The scientific study and debate about Gaia is an ongoing process, but one which should be distinguished from the metaphysical speculations often linked to it. The physicist Fritjof Capra welcomed the emergence of Gaia as a sign of a change of attitude, so that the Earth 'not just functions *like* an organism, but actually seems to be an organism... the new paradigm is ultimately spiritual'.[2] Gaia's proposer, Jim Lovelock, seems equivocal about this. He has written:

> For every letter I got about the science of Gaia [following his original pronouncement in a 1979 book] there were two concerning

---

[1] The golden rule: a proper scale for our environmental crisis. *Natural History*, **99**: 24–30, 1990.

[2] *The Turning Point: Science, Society, and the Rising Culture*. New York: Simon and Schuster, 1982.

religion. I think people need religion, and the notion of the Earth as a living planet is something to which they can obviously relate. At the least, Gaia may turn out to be the first religion to have a testable scientific theory embedded within it.

BOX 8.2 Gaia theory

The 'Gaia hypothesis' is that the Earth is a self-regulating, homeostatic system. It was put forward by British chemist Jim Lovelock following a request from NASA to devise a test for detecting life on Mars. He reasoned that the atmosphere of a lifeless planet must be in equilibrium with the physical composition of that planet, and hence would consist mainly of carbon dioxide, with a small amount of nitrogen and almost no oxygen. Such a planet would have a very high surface temperature due to the blanketing (or greenhouse) effect of the carbon dioxide. Any deviation away from this equilibrium situation would indicate the presence of a disturbing influence, which could be regarded as 'life'. As a chemist, he defined life somewhat exotically as 'a member of a class of phenomena which are open or continuous systems able to decrease their internal entropy at the expense of substances or free energy taken in from the environment and subsequent rejected in a degraded form…'. Mars turned out to have an atmosphere precisely as expected from its geological structure, but – and this was what started Lovelock thinking – the Earth's atmosphere is radically different from expectation.

The traditional assumption of the origin of life on Earth about 4000 million years ago is that it was the outcome of chemical processes involving progressive adaptation to contemporary atmospheric conditions which were originally reducing but then became oxidising. Lovelock turned these ideas upside down. His proposal was that the atmosphere changed in response to the life developing in it. In other words, that life (or the biosphere) regulates or maintains the climate and the atmospheric composition, and thus provides an optimum for itself. If this is true, the whole geobiochemical system can be regarded as a single, gigantic self-regulating unit. It was Lovelock's neighbour, William Golding (Nobel winner and author of *Lord of the Flies*), who suggested the name 'Gaia' for this system, after the Earth goddess of ancient Greece.

Gaia can be treated as a scientific or as a metaphysical theory. As the former, it has been a great success whether or not it is true, because of the research it has stimulated. Lovelock has claimed five successes for the theory:

1. prediction of the lifeless state of Mars made in 1968, confirmed in 1977 (although regularly questioned by astronomers);
2. carbon dioxide influence on climate through the biological weathering of rock predicted in 1981, shown in 1989;
3. the constant 21 per cent of oxygen in the atmosphere for the last 200 million years that could be due to fire and phosphorus cycling, for which the biological input is important;
4. the transfer of elements necessary for life on land, predicted in 1971 to be mediated through algae, shown in 1973; and
5. a link between cloud cover and the dimethyl sulphide produced by phytoplankton in the deep oceans – confirmed in 1987.

The most serious criticisms of the Gaia hypothesis have come from biologists. Richard Dawkins wrote in his 1982 book, *The Extended Phenotype*:

> Homeostatic adaptations in individual bodies evolve because individuals with improved homeostatic apparatuses pass on their genes more effectively than individuals with inferior homeostatic apparatuses. For the analogy [that the whole Earth is equivalent to a single living organism] to apply strictly, there would have to be a set of rival Gaias, presumably on different planets. Biospheres which did not develop efficient homeostatic regulation of their planetary atmospheres tended to go extinct. The Universe would have to be full of dead planets whose homeostatic regulation systems had failed, with, dotted around, a handful of successful well-regulated planets of which Earth is one.

Lovelock responded to this criticism by devising a computer model called Daisyworld which showed, he believed, that the regulatory behaviour which he postulated for Gaia could develop simply

> as a property of the complex processes which link organisms to their environment. There is certainly truth in the idea that living organisms (as defined in their normal as opposed to the Lovelockian way) modify their environment, and this can lead to natural selection.

Any change in the Earth's circumstances would be expected to result in an inevitable re-adjustment through natural feedback systems. Notwithstanding, Lovelock has had to accept that the Gaian system seems unable to cope with the stresses now being imposed on it, particularly from the vast amounts of greenhouse gases that we have been releasing into the atmosphere. Lovelock now speaks of 'the revenge of Gaia' on those who are assaulting it. Increased atmospheric carbon dioxide (CO_2) means that more is dissolved in the oceans, leading to them becoming more acidic, with implications for the productivity of the plankton on which the marine food chains depend.

BOX 8.3   **Climate change**

The gases in the atmosphere around the Earth serve as a blanket. Without it, the mean temperature would be about –19 °C, 33 °C lower than the current average. The chief of these 'greenhouse gases' is CO_2. As long ago as 1896, the Swedish chemist, Svante Arrhenius, calculated that doubling the concentration of CO_2 in the atmosphere would increase its blanketing effect, giving a mean global temperature rise of about 5 °C. This change may seem trivial, but it is actually about half the difference between the coldest part of an Ice Age and the following warm period. Past concentrations of atmospheric CO_2 can be measured from bubbles trapped in ice drilled from the Antarctic and Greenland ice caps. These concentrations remained more or less constant for at least 100,000 years until around 1750; they then began to increase, slowly at first, but then more rapidly. This continued (and seemingly inexorable) trend has been

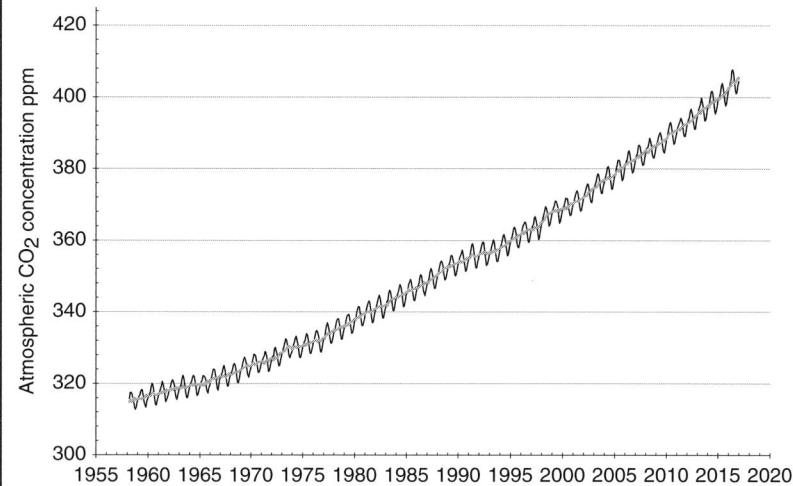

FIGURE 8.1 The seemingly inexorable increase in atmospheric CO_2 concentrations measured at the summit of Mauna Kea, an extinct Hawaiian volcano, chosen as a site as far as possible from any source of industry. Concentrations decrease in the northern summer months through uptake by photosynthesizing plants, but rebound in the winter. Initiated in 1958 by Charles Keeling of Scripps Oceanography and often called the 'Keeling Curve'.

monitored in detail since 1957 by measurements on the summit of Mauna Kea in Hawaii, chosen as a site as far as possible from industrial processes which generate CO_2 (Figure 8.1).

Extrapolation of the present rate of increase suggests that the mean global temperature might rise between 2 °C and 6 °C in the twenty-first century. There is a widespread assumption that a rise of more than 2 °C will cause major disruption to life on Earth, both human and non-human. The Earth Summit in Rio in 1992 agreed a framework Convention on Climate Change, with an objective 'to stabilise greenhouse gas concentrations in the atmosphere at a level that does not cause dangerous interference with the climate system'. Despite this, the Intergovernmental Panel on Climate Change (set up by the World Meteorological Organization and UN Environmental Programme (UNEP)) declared in 2014:

> The period from 1983 to 2012 was likely the warmest thirty-year period at the Earth's surface than any preceding decade of the last 1400 years...

> Anthropogenic greenhouse gas emissions have increased since the pre-industrial era, driven largely by economic and population growth... Their effects ... are extremely likely to have been the dominant cause of the observed warming since the mid-twentieth century.[1]

An analysis by the British economist Nicholas Stern concluded that without action to reduce the impact, 'climate change will reduce welfare by an amount equivalent to a reduction in consumption per head of between 5 and 20 per cent... [and] likely to be in the upper part of this range'.[2]

The 2003 heatwave in Europe is estimated to have caused damage to agriculture and forests of about £13 billion. Worse are the effects of climate change in communities least able to cope – in sub-Saharan Africa and south-eastern Asia, and on low-lying Pacific islands. By 2010, 30 per cent of the land in sub-Saharan Africa had become unsuitable for growing maize and 60 per cent was unsuitable for beans.

[1] *Climate Change 2014 Synthesis Report: Summary for Policymakers.* Geneva: Intergovernmental Panel on Climate Change, 2014.

[2] *Stern Review: The Economics of Climate Change.* Cambridge: Cambridge University Press, 2007.

Can we measure our impact on the Earth? One way would be to discover how much we take from it. We can calculate – very roughly indeed – the 'net primary production' of the Earth: that is the amount of organic material available (in the form of carbon compounds synthesized by photosynthesis) after subtracting the respiration of primary producers (mainly plants) from the total amount of energy (mainly from the Sun) that is fixed biologically. A very crude estimate of this global figure is 333 g a year on every square metre of the Earth, or just over 10^9 tonnes overall. This gives a baseline from which we can ask how much we take for ourselves.

We use some of this production directly – for fuel, fibre (such as cotton) or timber for building. Then we have to take into account the productivity of land devoted to human activities, most obviously for food. Putting these together with the misuse of land leading to desertification or degradation through overgrazing or erosion gives us as good an estimate as we can make of our impact.

Any calculation we make will be extremely rough, because all the figures we use are necessarily very approximate. Bearing this in mind:

- We eat about 1 million tonnes of plant growth a year.
- Our domestic animals eat about 2 million tonnes of plant growth a year.
- We use about 2 billion tonnes of plant growth each year for building materials and fuel.
- The global area of agricultural crops is 15 million km^2, producing 1700 tonnes of crops per km^2, a total production of 26 billion tonnes per year.
- We have built over land which would have generated 3 billion tonnes of growth per year.
- Forest converted to grazing land accounts for 3 billion tonnes of plant production per year.
- We harvest or burn 1500 tonnes of temperate and boreal trees per km^2 per year, a total of ¾ billion tonnes.
- Tree plantations make up 3 billion tonnes of biomass used per year.
- We clear 160,000 km^2 of tropical forest each year, but fires and selective logging damage several times more than this yearly.
- At least 5 million km^2 of tropical forest have been converted to pasture or otherwise changed.
- Salt accumulation from irrigation destroys the productivity of 15,000 km^2 of cropland per year and reduces the productivity of 450,000 km^2.
- Erosion and other agricultural practices in arid regions destroy about 20,000 km^2 each year and reduce the productivity of another 2 million km^2.
- About 6 million km^2 of pasture lands are burnt each year.
- Overgrazing has damaged productivity over 35 million km^2 of drier lands.

Putting all these data together, the answer is that our activities account for 60 billion tonnes of actual or potential primary productivity per year: we harvest 26 billion tonnes directly from agricultural crops and 14 billion from forests, but forego 3 billion because of land loss through buildings and another 17 billion from grazing. The bottom line is that we are using over 40 per cent of the terrestrial net primary production.

Similar, although even less precise, calculations can be made for fisheries. Overall, we are using 35 per cent of the productivity of the continental shelves – the most fertile areas of the seas. In addition we use 60 per cent of available fresh water. When we add to this our taking of the 40 per cent of annual terrestrial plant growth, we cannot avoid the conclusion that we have a very major impact on the Earth – and this rate of extraction has doubled in the past century. The conclusion is further complicated by the continuing increase in human numbers – conservatively estimated to be around 50 per cent within the next century. And on top of all this, we face certain but unquantified changes in climate during the same period.

All this means very little to most people. Global budgeting on this scale is well outside our normal horizons, but it carries the clear implication that we are running out of world. In one way, there is nothing new about this: it has been happening on a regional scale throughout history. The traditional response to overpopulation and land scarcity has been to move – the Beaker Folk, Teutons, Vikings and New World colonizers all spilled from the western seaboard of Europe. More recently, increasing numbers of Africans have been seeking to uproot to Europe. On top of this, mismanagement has often produced disastrous consequences for humans: introduction of rabbits and cane toads to Australia, mongooses to Hawaii and grey squirrels to Britain have all caused major ecological problems. The difference now is that the problems are global, not regional; we have nowhere left to flee.

BOX 8.4 **Ecological footprinting**

Ecological footprinting is another way of measuring human demand on natural resources, with the advantage that it can identify the contributions of particular individuals or communities. It involves estimating the area of land used to supply an individual

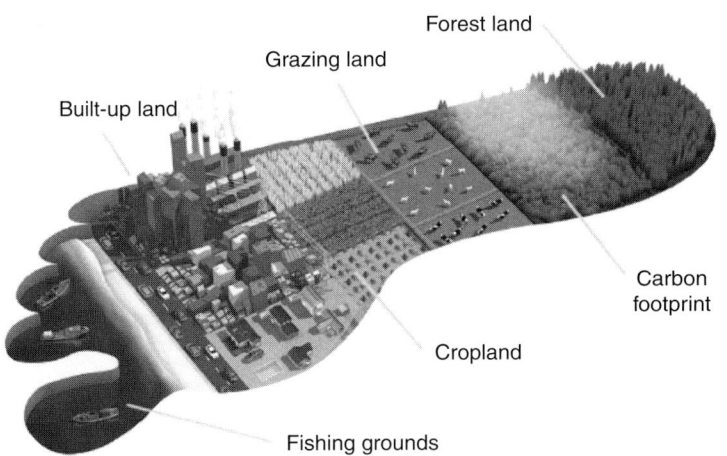

Forest land

Grazing land

Built-up land

Carbon footprint

Cropland

Fishing grounds

FIGURE 8.2A Our ecological footprint is the notional area from which we (individually or as a community) take from the environment for food, fuel, recreation, disposal of waste (including CO_2), living space, etc.
© 2016 Global Footprint Network. www.footprintnetwork.org.

or community and to deal with the waste produced. It provides a method of comparing consumption and lifestyles (Figure 8.2).

Globally, the average productive area used per person is 1.8 ha (just over three football pitches), but this varies enormously between 10.7 ha in the United Arab Emirates, 9.0 ha in the USA, around 5 ha in the UK and much of Europe, to less than 1.0 ha in India, Pakistan and sub-Saharan Africa. The unhappy conclusion is that we are consuming 2.7 ha each, or about 30 per cent more than our 'ecological income' (i.e. the productivity of our 1.8 ha share). This means that we are living off 'ecological capital', most obviously in our consumption of fossil fuels. Indeed, a parallel concept to 'ecological footprint' is 'carbon footprint', which is the amount of CO_2 and methane (so-called 'greenhouse gases') produced by an individual or community. This can be broken down in various ways, conventionally by industrial, domestic and agricultural emissions and travel.

The concept of footprints is linked to the notion of 'Spaceship Earth', to express concern that there are finite and therefore limited

FIGURE 8.2B In many societies the demands of our lifestyle mean that our footprint exceeds the natural (or renewable) income of the environment; in other words, we are using up natural capital, using resources unsustainably.
Reproduced with permission from Rees, William E. and Wackernagel M. (1997). Our ecological footprint: where on Earth is the Lower Fraser Basin? In Healy, M.C. (ed.). *Sustainability Issues and Choices in the Lower Fraser Basin: Resolving the Dissonance*. Vancouver: University of British Columbia Press.

resources available on Earth (p. 185). In *The Road to Wigan Pier* (1937), George Orwell wrote:

> The world is a raft sailing through space with, potentially, plenty of provisions for everybody; the idea that we must all cooperate and see to it that everyone does his fair share of the work and gets his fair share of the provisions seems so blatantly obvious that one would say that no one could possibly fail to accept it unless he had some corrupt motive for clinging to the present system.

The notion has been used in various guises since then, such as the concept that 'small is beautiful' in a book of that name by Ernst Schumacher (1973); it obviously relates to a range of other themes, including sustainability and *The Limits to Growth* (p. 189).

Despite all this, we rarely accept the implications. We have repeatedly seized upon some new crop or technique as a golden bullet which will solve a host of problems at a stroke. Subsidies have been used to encourage production – too often leading to distortions such as wine lakes or butter mountains. In recent years there was a spurt of enthusiasm for growing crops which could be processed into (mainly) ethanol as a biofuel – but the amount of land diverted for this endeavour led in some places to a lack of agricultural land for food crops. Palm oil has been seen as a major contributor to sustainability: for human consumption, sequestering atmospheric carbon or for processing into biodiesel. Between 2006 and 2010, palm-oil plantations in Indonesia increased from 4.1 million to 7.2 million ha – but it involved massive deforestation, water and air pollution and soil erosion.

These sorts of problems highlight the need for environmental literacy. Literacy involves the ability to read, but also requires comprehension. A literate person is one who can understand what is written and is able to place it within a context. 'Environmental literacy' involves 'head learning' about the environment from books and libraries, but goes further and involves the experiencing and interpreting of real environments. This was the inspiration that Charles Darwin and John Muir got from their travels. It provided the stimulus that the awe of the natural world evoked. But this perception has now become much broader and more complicated. We have, on the one hand, the often reproduced image of the Earth floating in space ('the only planet we have'), emphasizing our finiteness and the reality of our limited resources – magnified by instant television pictures of degraded landscapes which reinforce our frailty and insignificance; on the other, common observation of vastness – particularly from long-distance plane journeys – showing us that the world is very big compared to our own restricted perception and suggesting that our behaviour might be so puny as to have no material effects on the world as a whole. We know the facts but dismiss their possible implications as scaremongering.

We only too often disagree about how to put together these sorts of observations. We almost routinely exhibit a failure of ecological literacy which is almost culpable. In the 1960s, the debate was between proponents of 'zero population growth' of human populations and those who saw salvation coming from better and more efficiently used resources. These issues should have been faced much earlier. Two and a half centuries ago (in 1758) Otto Lütken, a Danish pastor, wrote in the *Danish Norwegian Economic Magazine*:

> Since the circumference of the globe is given and does not expand with the increased number of its inhabitants, and as travels to other planets thought to be habitable has not yet been invented; since the earth's fertility cannot be extended beyond a given point and since human nature will presumably remain unchanged, so that a given number will hereafter require the same quantity of the fruits of the earth for their support as now, and as their rations cannot be arbitrarily reduced, it follows that the proposition 'that the world's inhabitants will be happier, the greater their number' cannot be maintained, for as soon as the number exceeds that which our planet with all its wealth of land and water can support, they must necessarily starve one another out, not to mention other necessarily attendant inconveniences, to wit, a lack of other comforts of life, wool, flax, timber, fuel and so on. But the wise Creator who commanded men in the beginning to be fruitful and multiply, did not intend, since He set limits to their habitation and sustenance, that multiplication should continue without limit.

FURTHER READING

Houghton, J. (2015). *Global Warming: the Complete Briefing*, 4th edition. Cambridge: Cambridge University Press.

Lovelock, J.E. (1979). *Gaia: a New Look at Life on Earth*. Oxford: Oxford University Press.

Pimm, S.L. (2001). *The World According to Pimm*. New York: McGraw-Hill.

9 Running Out of World

The most dangerous worldview is the worldview of those who have not viewed the world.

Alexander von Humboldt

Lütken's warning could be regarded as a challenge to individuals, but more importantly it shows the snowballing of environmental use as numbers increase. Family size is a matter for individuals to decide, but choices made without consideration of global consequences have the danger of leading to local overpopulation and overexploitation of resources. This can too easily scale up to nations and even civilizations. These considerations therefore need to be on the agenda of national and international bodies, with the result that we often feel like mere pawns in events well beyond our control. Worse, the efforts to deal with this may itself lead to disaster, as putative solutions produce complex layers of bureaucracy.[1] This resulting impotence has been all too graphically and rather gruesomely spelt out by Jared Diamond in a book called *Collapse*, published in 2005.

Diamond examined the biological factors which contributed to the fate of a range of once flourishing cultures: the Anasazi in New Mexico, the Mayans of Central America, the Moche and Tiwanaku societies of eastern South America, Mycenean Greece, Minoan Crete, Greater Zimbabwe, Angkor Wat – and warned of the dangers faced by many contemporary cultures if we fail to heed lessons from them. There were many reasons for the disasters described by Diamond, but he extracted five common underlying and interacting

[1] Tainter, J.A. (1988). *The Collapse of Complex Societies.* Cambridge: Cambridge University Press.

causes: environmental damage, climate change, hostile neighbours, trade partners and how those affected reacted to symptoms of stress. The details of his analyses have been challenged, but the main conclusions are firm. Some of the causes were beyond the ability to respond of those affected – such as climate change – but others might reasonably have been foreseen and acted on: habitat destruction (such as deforestation and desertification), soil degradation (erosion, water-logging, salination), water supply problems, overhunting, overfishing, the impacts of introduced species, population growth pressures and rising per capita impacts. Overall, Diamond concluded that environmental degradation dominated, but the final and decisive factor in all the situations he examined was a blindness to future consequences – a failure to read trends and see behind the phenomenon of 'creeping normalcy'; things getting just a little bit worse year by year but never bad enough for anyone to take action. Superimposed on all this was the disproportionate power of conventionally reactionary elites, particularly when they condone or even positively promote what he describes as 'rational bad behaviour' on the part of those who manage or use natural resources. Two civilizations which lasted longer than most (in Egypt and China) survived largely because of the lavish subsidies of fresh topsoil brought in regularly from elsewhere by wind (China) or water (Egypt).

The societies Diamond examined did not collapse overnight, but suffered decline and transformation over long periods. The insidious nature of such declines was more dangerous than a single disaster. Alarm bells about the danger of chemical pesticides used on roadside verges were first raised in Britain in 1950. The damage caused on non-target species was confirmed in a series of studies. In 1958, twenty cows died after feeding on insecticide-treated weeds. Bureaucrats began to accept that there was a problem needing investigation. This led in 1960 to a 'Toxic Chemicals and Wild Life Section' being set up in the Nature Conservancy's Experimental Station. Over the next few years it produced significant results, showing the accumulation of persistent chemicals in food chains, perhaps most notably the thinning of bird egg shells, which resulted in the massive decline of peregrine falcons and other raptors.[2] However, the key stimulus to action about chemicals came from the USA with the publication of Rachel Carson's *Silent Spring* in 1962. It highlighted the consequence of gratuitously ignoring processes which had been creeping up unnoticed. In his introduction to the British edition of *Silent Spring*, Lord Shackleton wrote of 'a feeling of a lack of urgency... The agricultural establishment finds it difficult to see the wider and longer-term consequences... Nothing like enough research is being done.' Julian Huxley added a preface, noting that 'Ecology in the service of man... must aim not only at optimum use but also at optimum conservation of resources [which] include enjoyment resources like scenery and solitude, beauty and interest, as well as material resources like food or minerals.'

BOX 9.1   *Silent Spring*

The author of *Silent Spring*, Rachel Carson, was a marine biologist employed as a science writer. She was a keen naturalist who retired to her house on the coast of Maine whenever she could, and was sensitive to her immediate environment and changes in it. A mass death of birds in Cape Cod following spraying of the insecticide DDT to control mosquitoes concerned and alerted her, and she began to enquire about the effects of the wholesale use of chemicals designed for pest and disease control, and its possible consequences for humans. She portrayed the change from a community where all life seemed to live in harmony with its surroundings to one where only silence lay over the fields and woods and marsh. She backed up her gloom with a formidable mass of data and wrote about it in a simple and dramatic way, albeit with relevant statistics and involving a call to action. She was not the first person to warn about the indiscriminate use of poisons. As noted above, questions about their use were being raised in Britain, and in the USA Murray Bookchin had written about potentially dangerous chemicals in human foods,

---

[2] Sheail, J. (1985). *Pesticides and Nature Conservation*. Oxford: Clarendon Press.

ten years before *Silent Spring*, but he was the archetype of a radical 'green' and his alerts were not taken seriously.

Silent Spring was published at the height of the Cold War, when science and technology were seen as essential allies in defeating the USA's enemies, and in a time when cereal output had increased threefold as a result of new genetic varieties and the introduction of chemical fertilizers and pesticides. The paradox between the gains from science and the dangers from its misuse increased its impact; it inspired many groups, pieces of legislation and government agencies. It also provoked furious resistance, chiefly from the big chemical companies. They argued that Carson had failed to consider the benefits of pesticides; they claimed that taking her advice to reduce insecticide use would lead to the return of the scourges of famine and poverty which had been almost abolished. But Carson's analysis held up, and slowly the government responded, banning the most damaging organochlorine pesticides by the 1980s.

The fact that Carson wrote in a way that the general public could understand rather than for a scientific or political elite undoubtedly increased the book's impact. Attempts were made to destroy not only Carson's scientific credibility but also her character: she was condemned as a fanatic, a 'tree hugger', a dangerous reactionary who would drag modern society into a new Middle Ages filled with pests, vermin, crop destruction and lethal diseases. These attacks turned out to be counter-productive. Carson proved convincing.

The book raised awareness of the risks to both humans and non-target species; it led to action in the form of legislation to regulate pesticide use. In 1968, husband and wife team Paul and Anne Ehrlich, in a widely read book *The Population Bomb*, drew attention to another inexorable pressure on the planet: the exponential increase in human population. In the USA some of the functions of the Department of Agriculture (USDA) were transferred in 1970 to a newly formed Environmental Protection Agency. Two years later the first truly international conference on these issues was held – the UN Conference on the Human Environment in Stockholm (p. 190). A lasting consequence of the Stockholm Conference was the setting up of UNEP.

Silent Spring appeared in 1962. Together with images of the Earth in space which began to appear in 1972, it served as a signal, preparing the way for a more inclusive environmental understanding and response, a foretaste of the complexity of issues shown by Jared Diamond's analyses.

The 1960s were significant for environmental awareness. The Countryside in 1970 conferences in Britain (p. 162) proved to be harbingers of bigger and lasting events. In 1966, British-born economist Kenneth Boulding took up Orwell's analogy, pointing out that we are living in 'a spaceman economy, in which the Earth has become a single spaceship, without unlimited reservoirs of anything, either for extraction or pollution'; three years later, Buckminster Fuller emphasized this with an *Operating Manual for Spaceship Earth*, warning of the consequences for us of the finiteness of the world and its resources. Then in April 1970, thirty individuals from ten countries met at the instigation of Aurelio Peccei, a visionary Italian industrialist, in the Accademia dei Lincei in Rome. This 'Club of Rome' commissioned a study on 'the Predicament of Mankind'. Another group based at the Massachusetts Institute of Technology (MIT) had similar concerns and produced a report on *Man's Impact on the Global Environment* (1970), which showed unequivocal evidence of specific pollutant damage in the atmosphere, ocean ecology and large terrestrial ecosystems. The Swedish Government had similar concerns and was keen for international debate on human interactions. All this fed into the UN Conference on the Human Environment in Stockholm in 1972.

Three months before the Conference convened, the first fruits of the Rome meeting were published, a study called *The Limits to Growth*. It was a computer simulation carried out at MIT on the effects of the use (and inevitable depletion) of non-renewable resources. Its message was very simple: that the economic and industrial systems of developed countries would collapse around the year 2100 unless birth and death rates became the same (in other words, that population numbers stabilized) and capital investment matched capital depreciation. It marked the beginning of increasingly complex computer-based modelling and the production of scenarios linking (indeed, requiring) sustainability to the future of humanity.

The *Limits to Growth* study was savaged by economists on the grounds that it ignored market forces and technological developments. They argued that when a resource became scarce, its price increased or it was replaced by a cheaper substitute. Notwithstanding, it resonated widely with the lay public by drawing attention to the fact that the world is a finite system with fixed boundaries, even if we cannot agree what these are or when we shall reach them. It was a message which became even more stark when the Yom Kippur War exploded fifteen months later, bringing with it an oil crisis. Follow-up re-analyses twenty and thirty years after the original report using better data and more refined programmes have merely confirmed the initial conclusions.

The *Limits to Growth* study drew attention to the danger of ignoring our impact; it showed that our survival depends ultimately on our rate of use of resources and whether the resources we use are renewable or not. It should have produced realism; in fact, it induced despair rather than determination. It was assumed to point to the inevitability of 'doom'. Notwithstanding, it complemented the message of the Stockholm Conference that 'A point has been reached in history when we must shape our actions throughout the world with a more prudent care for their environmental consequences. Through ignorance or indifference we can do massive and irreversible harm to the earthly environment on which our life and well-being depend.'[3] A basic concept of the Conference was 'development without destruction'. The starting gun for global action had been fired.

The Stockholm Conference set out a series of 'principles', one of which was that 'Economic and social development is essential for ensuring a favourable living and working environment for man and for

[3] *Declaration of the United Nations Conference on the Human Environment.* New York: United Nations, 1972.

creating conditions on earth that are necessary for the improvement of the quality of life.' This had the unfortunate effect of apparently separating social and economic development from environmental conservation. Eliminating poverty was obviously a prime concern (and an absolute priority for developing nations). Conservation came to be regarded as an indulgence for developed nations, acceptable only when the basic requirements of poverty alleviation, adequate housing and sanitation were achieved. This assumption strengthened through the 1970s, not helped, at the 1982 Nairobi conference marking ten years since Stockholm, by the way in which the UNEP – which had been one of the positive outcomes of the Stockholm Conference – was marginalized, far from other main sites of international action. Taking care of the environment was seen by many developing nations as likely to hold back their own ambitions for development; environmental care was regarded as a much lower priority than attacking poverty, and therefore a hindrance to development.

The split between development and environment was far from the policies, attitudes and practices of the Stockholm Conference, but it became ever clearer that there was little hope of public leaders recognizing this. National and international organizations were split among different interests – agriculture, forestry, fisheries, wildlife. UNEP found it difficult to provide leadership. However, a way forward emerged – a collaboration between UNEP and IUCN.

IUCN had been mulling over the possibility of a strategy document on conservation for some years. In 1975 Duncan Poore (Secretary-General of IUCN and a former Director of the British Nature Conservancy) and Robert Prescott-Allen (of the IUCN staff) had begun work on a World Conservation Strategy. This was formally published in 1980, jointly with UNEP. Peter Scott wrote that:

> it represented a change in attitude. The confident assumption of the 1950s and 1960s that man would find solutions to all his problems has been supplanted by a new humility, born of the realization that even man's most astonishing achievements cannot offset his

disastrous devastation of the earth, its plants and animals... We must recognize that we are a part of nature and must resolve that all our actions take this into account.

At the launch, the Indian Prime Minister, Indira Gandhi, rejected the term 'Conservation Strategy'; for her, it was a 'survival strategy' for the growing billions of India. The naturalist Richard Mabey described it as 'the first major international policy statement to affirm the *human* importance of nature conservation... it is vital that we recognize that deliberately intervening or refraining from intervening, or adopting any conscious positive strategy towards the natural world is a human *choice*, whether it is for reasons of economic self-interest, scientific inquisitiveness, faith or pure sentimentality'.

The Strategy combined the sensitivity of UNEP to developmental concerns while going far beyond the original preoccupation of IUCN with nature and wildlife. A key intention was to counter the assumption that conservation and development could be separated. Because of its wide implications it became a blueprint adopted by both UNESCO and the Food and Agriculture Organization.

The World Conservation Strategy had three explicit aims:

- to maintain essential ecological processes and life-support systems;
- to preserve genetic diversity; and, significantly
- to ensure the sustainable utilization of species and ecosystems.

As Mabey recognized, the Strategy focused firmly on people: 'Humanity's relationship with the biosphere (the thin covering of the planet that contains and sustains life) will continue to deteriorate until a new international economic order is achieved, a new environmental ethic adopted, human populations stabilize, and sustainable modes of development become the rule rather than the exception.'[4] Sustainability became part of the international vocabulary.

---

[4] *World Conservation Strategy: Living Resource Conservation for Sustainable Development.* Gland: IUCN, 1980.

The implications of all this were taken up by a World Commission on Environment and Development chaired by Gro Harlem Brundtland, whose Report in 1987 (*Our Common Future* – commonly known as the Brundtland Report) urged the need to recognize ecological as well as economic interdependence among nations. The working title of the Report was *Our Common Future?* but the Commission members were confident that there could be a future for humanity and removed the question mark. Brundtland steered the group away from dealing only with environmental issues; she underscored the principle that separating environment and development was illegitimate. The Report linked economic, social, environmental and cultural sustainability. This comprehensiveness has been almost forgotten; the Report tends to be remembered only for its definition of sustainable development: 'Development that meets the needs of the present without compromising the ability of future generations to meet their own needs'. Although oft-quoted, this definition has been heavily criticized as ambiguous and open to contradictory interpretations. The problem is that it could be read as ignoring the constraint of 'limits to growth' (p. 189), with the implication that 'nature' has the capacity to meet all human needs once social and technological deficiencies are sorted out. This belief still persists in some places, perhaps most commonly among politicians who see the obvious way forward is to deal with deficiencies rather than the underlying constraints.

The Stockholm Conference, the World Conservation Strategy and the Brundtland Commission all called upon and urged integrating a range of disciplines. They represented a wide international concern about the environment, but also a move even further away from individual involvement. In 1989 the Economic Summit Nations (the G7) held a conference on environmental ethics. European Commission President Jacques Delors opened with a description of the problems he believed must modify our approach and behaviour towards the environment:

> None of these problems [climate change, the depletion of biological diversity, the growing depletion of resources] can be approached

separately: they mesh together and transcend our traditional frameworks of thought and action… [They] highlight man's [*sic*] dependence on his environment, hitherto ill-perceived. They underscore the sudden fragility of man's relationship with nature, which has traditionally been one of mastery based on use and exploitation. It is in the broadest sense that the very conditions of humanity which current problems compel us to rethink and rebuild, insofar as continuing our traditional modes of life on earth would lead to ever-increasing damage and before long threaten to destroy us.

Based on this diagnosis, Delors saw the way forward as needing an ethical approach, an approach that:

concerns the values which govern social behaviour. It is the bedrock of law and therefore determines the various codes by which we act, codes hallowed by tradition and whose real cruxes must now be re-established. The constant degradation of the setting for life which man had received from his forebears will of necessity prompt him to adopt an approach to that legacy in terms of duties and responsibilities.

Prompted by this, the conference produced a deceptively simple statement for an acceptable environmental ethic: 'Stewardship of the living and non-living systems of the earth in order to maintain their sustainability for present and future, allowing development with forbearance and equity. Health and quality of life for humankind are ultimately dependent on this.'[5] It used this as a basis for a 'Code of Environmental Practice' which was endorsed by the Heads of State of the G7 Nations at their next meeting – although immediately forgotten. Nevertheless, it was a step along the way of an emerging international consensus on the environment.

[5] Appendix A: a code of environmental practice. In: Berry, R.J. (ed.). *Environmental Dilemmas: Ethics and Decisions*. London: Chapman & Hall, 1992.

In 1990 the Secretary-General of the UN, Javier Perez de Cuellar, called in his Annual Report for concerted action on the environment: 'The Charter of the United Nations governs relations between States. The Universal Declaration of Human Rights pertains to relationships between the State and the individual. The time has come to devise a covenant regulating relations between humankind and nature.' This challenge was taken up by the International Environmental Law Commission with a 'Covenant on Environment and Development', intended as a complement to the Declaration on Human Rights. It brought together the Stockholm Conference Declaration, the World Charter for Nature (accepted by the UN in 1982), a mass of other documentation and 'soft law', but most particularly understandings and pressures voiced at the Earth Summit in Rio in 1992, in which 172 governments participated.

Delors's call for ethical issues to be taken seriously was represented in a revised World Conservation Strategy, which formed the base document for the Earth Summit. The authors of the original Strategy had brain-stormed its impact and weaknesses.[6] They identified a significant problem as being that the Strategy had fallen into the common Enlightenment fallacy of failing to emphasize that understanding did not automatically produce action – that responsible behaviour towards the environment was not an inevitable result of awareness of environmental facts. It became clear that there was a need to stress the links between science and ethics, and to bind social justice with ecological wholeness. This prompted IUCN (the main author of the Strategy) to set up an Ethics Working Group to feed into the drafting of the revised Strategy (published as *Caring for the Earth*, 1991).

Caring for the Earth called for 'a world ethic for living sustainably'. It also set out a re-definition of 'sustainable development' as *improving the quality of human life while living within the*

---

[6] Jacobs, P. and Munro, D.A. (eds.). (1987). *Conservation with Equity*. Gland: IUCN.

carrying capacity of supporting ecosystems. It pointed out the confusion around the uses of 'sustainable' as an adjective:

> 'Sustainable development', 'sustainable growth', and 'sustainable use' have been used interchangeably as if their meanings were the same. They are not. 'Sustainable growth' is a contradiction in terms: nothing physical can grow indefinitely. 'Sustainable use' is applicable only to renewable resources: it means using them at rates within their capacity for renewal... A 'sustainable economy' is the product of sustainable development. It maintains its natural resource base. It can continue to develop by adapting, and through improvements in knowledge, organization, technical efficiency and wisdom.

BOX 9.2   **Carrying capacity**

'Carrying capacity' is the number of individuals which can be supported indefinitely (or sustainably) in a particular habitat. It is a concept commonly used by ecologists and frequently misused by politicians. The problem is that it depends on a complex of constraints. If a population of animals has no limitations on food or space, it will grow exponentially in numbers, much as the global human population has been doing, at least until recently. However, assumptions of unlimited food and space are unrealistic. As numbers increase, factors both intrinsic and extrinsic will begin to affect the rate of population growth until birth and death rates become equal and the population size stabilizes. This balance is the carrying capacity of that population in that place. Attaining this balance reflects Thomas Malthus's nightmare as he viewed possible food shortages in the emerging nineteenth-century world and reacted against the unwarranted optimism – as he saw it – of the utopian speculation of people like Robert Wallace (*Various Prospects of Mankind, Nature, and Providence*, 1761) and William Godwin (*Enquiry Concerning Political Justice*, 1793).

It is wrong to refer to *the* carrying capacity of the environment without specifying the conditions that determine it. Joel Cohen has listed sixty-five different estimates of the Earth's total carrying capacity.[1] He separates those which focus on a single assumed constraint (usually food) from those which recognize that 'men and women do not live by bread alone. People also require wood, fibre, fuel and amenities.' These needs can be broken down into physical (shelter, food, clean air, water), economic (transport, shops, work) and aesthetic (space, quiet, access to countryside).

[1] *How Many People can the Earth Support?* New York: W.W. Norton, 1995.

What is 'quality of life'? *Caring for the Earth* describes quality of life as:

> A process that enables human beings to realize their potential, build self-confidence and lead lives of dignity and fulfilment. Economic growth is an important component of development, but it cannot be a goal in itself, nor can it go on indefinitely. Although people differ in their goals that they would set for development, some are virtually universal. These include a long and healthy life, education, access to the resources needed for a decent standard of living, political freedom, guaranteed human rights, and freedom from violence.

Taken in isolation, this description could be taken from a self-help manual, but *Caring for the Earth* qualifies it with the ethic missing from the original World Conservation Strategy. A major aim was that everyone should:

> Share fairly the benefits and costs of resource use, among different communities and interest groups, among regions that are poor and those that are affluent, and between present and future generations. Each generation should leave to the future a world that is at least as diverse and productive as the one it inherited. Development of

one society or generation should not limit the opportunities of other societies or generations.

BOX 9.3   **Human well-being**

What is human well-being? Is it the same thing as 'quality of life', which includes economic factors but extends to health, education, recreation and leisure, and social belonging? While a basic income is necessary, many studies have shown that 'happiness' and 'contentment' do not increase at the same rate with income. There seems to be general acceptance that well-being includes contributions from nature; from family, friends and other social relationships; and from health, education and fulfilling employment. We depend on all these inputs, but nature is not merely an optional or luxury good.

Is it realistic to define human well-being? It is clearly more than economic affluence. In an analysis of the responses of 60,000 poor people in twenty-three countries, the World Bank identified five linked elements as the perceived components of well-being:

- the material provisions for a good life – including secure and adequate income and assets, enough food at all times, shelter, furniture, clothing, access to goods;
- health – including being strong, feeling well and having a healthy physical environment;
- good social relations – including social cohesion, mutual respect, good gender and family relations, and the ability to help others and provide for children;
- security – including secure access to natural and other resources, safety for oneself and one's possessions, life in a predictable and controllable environment; and
- freedom and choice – including having control over what happens and being able to achieve.[1]

The human-welfare-producing system depends on five different sorts of capital: financial, human, built, social and natural. These are also interdependent. Recognizing this, the British Government

---

in 1999 produced its strategy for sustainable development. It had four aims:

- social progress which recognizes the needs of everyone;
- effective protection of the environment;
- prudent use of natural resources; and
- maintenance of high and stable levels of economic growth and employment.

To measure progress towards these aims the Government identified 147 'quality of life' indicators, with fifteen of them regarded as 'headline indicators' (economic growth, investment, employment, poverty, education, health, housing, crime, greenhouse gas emissions, air quality, road traffic, river water quality, wildlife, land use and waste). It publishes progress towards (or away from) these each year. In general, the trend has been positive, although uneven.

These five World Bank elements are taken as the basis for measuring environmental interactions with humankind in such international exercises as the Millennium Ecosystem Assessment (p. 200).

[1] Narayan, D., Patel, R., Schafft, K., Rademacher, A. and Koch-Schulte, S. *Voices of the Poor: Can Anyone Hear Us?* New York: Oxford University Press, 2000.

Caring for the Earth set out the problems and possible ways forward for the Earth Summit. Whatever one thinks about the blandness of international conferences, the Earth Summit has had lasting effects. It agreed a 'Rio Declaration' (listing twenty-seven principles intended as a guide towards sustainable development), Agenda 21 (setting out an 'action plan' for the twenty-first century), and two Conventions – on Climate Change and on Biological Diversity. The Convention on Climate Change is the better known of the two because of its subsequent history and continuing controversy about the way to implement it. It recognized that climate change is a serious problem; that action cannot wait upon the resolution of scientific uncertainties; that the developed countries should take a lead in action; and that the developed countries should compensate developing countries for any additional costs incurred on taking measures under the Convention.

The Convention on Biological Diversity was designed to preserve the biological richness of the planet through protecting species and ecosystems, and establishing rules for the uses of biological sources and technology. It affirmed that states have 'sovereign rights' over biological resources in their territory, the fruits of which should be shared in a 'fair and equitable' way on 'mutually agreed terms'. One hundred and ninety-three states signed up to it. The only nation which has so far positively refused to do so is the USA, on the grounds that it imposed a restrictive regulatory framework on prospecting for organisms potentially contributing valuable genes and a failure to protect intellectual property rights.

It soon became clear that real data were needed to support the Convention on Biological Diversity. There were no relevant international agencies like the Intergovernmental Panel on Climate Change, set up in 1988 by UNEP and the World Meteorological Organization. There was an obvious requirement for some sort of global ecosystem assessment. How can one define or defend one's attitude unless one knows what is 'out there'? Although major advances were made in ecological sciences and resource economics during the 1980s and 1990s, they were poorly reflected in policy discussions concerning ecosystems. This lack led to a call by UNEP and the World Bank for 'a more integrative assessment process for selected scientific issues, a process that can highlight the linkages between questions relevant to climate, biodiversity, desertification, and forest issues'. The result was a 'Millennium Ecosystem Assessment' (MA), carried out by ecologists throughout the world to measure the consequences of ecosystem change for human well-being.[7] Its intention was to produce a basis for the conservation and sustainable use of ecological systems and their contribution to human well-being. Its report provided a state-of-the-art scientific appraisal of the condition and trends in

[7] https://www.millenniumassessment.org/en/

the world's ecosystems and the services they provide (such as clean water, food, forest products, flood control, and natural resources) and the options to restore, conserve or enhance the sustainable use of ecosystems.

The unsurprising bottom line of the MA findings was that human actions are depleting Earth's natural capital, and this is putting a strain on the environment to the extent that the ability of the planet's ecosystems to sustain future generations can no longer be taken for granted. On a positive note, the MA argued that appropriate actions should make it possible to reverse the degradation of many ecosystem services over the next fifty years, although the changes in policy and practice required would be substantial and were not currently under way.

BOX 9.4 **Ecosystem services**

The concept of ecosystem services first surfaced in a 1970 report, *Man's Impact on the Global Environment*, prepared for the Stockholm Conference. The MA defined ecosystem services as 'the benefits people obtain from ecosystems', and recognized four categories (not dissimilar to the criteria for sustainable development laid down by the British Government (p. 199)):

- supporting, i.e. systems necessary for the operation of other systems – nutrient recycling, primary productivity, soil formation;
- provisioning, i.e. products obtained from the natural world – food (for humans and domestic animals), building materials, fuel, fertilizer, genetic resources, water, minerals, energy (fossil fuels, plus wind, water, tidal power, biomass);
- regulating: carbon sequestration and other aspects of climate regulation, decomposition and detoxification of waste, water and air purification, pest control;
- cultural: 'nonmaterial benefits through spiritual enrichment, cognitive development, recreation, and aesthetic experiences'.

The concept has become increasingly prominent as the value of such services to humankind has been calculated in monetary terms. A very crude estimate of their overall magnitude is that the world's ecosystems add up to around $33 million million (trillion) every year to the world's sustenance, a sum twice the annual gross national product of all the nations of the globe put together. Ecosystem services operate on scales varying from the microbial to landscape, and from milliseconds to millions of years. They include, for example, the detritus on a forest floor, the microorganisms in the soil and the characteristics of the soil itself, all of which contribute to the abilities of the forest to influence carbon sequestration, water purification and erosion prevention.

Ecosystem services are only a part of the overall economic scene, but they are an essential part of it. We derive ecosystem services from the world's stocks of natural assets – geology, soil, air, water and living things. Together these constitute 'natural capital', which is an integral part of all the capital available to us: financial, manufactured, human, social and natural. Poorly managed natural capital will affect other forms of capital. Decline in (or 'spending') natural capital potentially threatens human well-being, understood rather imprecisely as a dynamic state of health, happiness and/or prosperity. Put another way, it gives people a sense of how their lives are going by highlighting the interaction between their circumstances, activities and psychological resources (or 'mental capital'). For example, half the world's topsoil has been lost in the last 150 years, affecting stability and agricultural productivity. Unsustainable land management costs between £4 trillion and £12 trillion each year. Well-being and deprivation can be regarded as different sides of the same coin.

UNEP has tried to refine the value of ecosystem services in a global programme (The Economics of Ecosystems and Biodiversity; TEEB) aimed at helping 'decision-makers recognize the wide range of benefits provided by ecosystems and biodiversity and demonstrate their values in economic terms'. Its main programme in Britain was a National Ecosystem Assessment, carried out between 2009 and 2011, building on the global MA. It found that Britain's forests are 'worth' fifteen times their value as timber because of the enjoyment they provide, the pollution they filter and the carbon they store. The stock of natural capital in Britain was estimated to be at least £1.6 trillion. It also recorded that the land used for growing crops increased 40 per cent between 1940 and 1980, with productivity increasing even more sharply (the yield of wheat per hectare grew threefold) as a result of national policy and agricultural improvements. In contrast, landings of fish and other seafood halved between 1970 and 2000. Examples of the monetary value of ecosystems include:

- Pollination of agricultural crops by insects – this would otherwise have to be carried out by hand. It is estimated that insect pollination is worth £250 billion to global agricultural output.
- Flood control – it is now recognized that this is often better achieved by controlling catchments than by expensive flood defences in built-up areas. The estimated annual benefits in the Tamar catchment on the Devon–Cornwall border from increased food production, climate regulation, recreation and tourism is £3.8 million, compared with the gross cost of maintaining it of £600,700.
- Soil degradation in the UK costs an estimated billion pounds a year through loss of organic matter, compaction and erosion. This can be reduced by controlling field drainage, rotation of crops and grazing, planting shelter belts, etc.
- Anaerobic digestion of agricultural and food wastes can produce energy on a large scale. There are over 200 anaerobic digester plants now operating in Britain. A typical plant is one near St Asaph, which annually processes 22,500 tonnes of food waste and generates a megawatt of electricity for local homes by feeding into the National Grid.
- Retaining hedges and permitting uncultivated 'headlands' around planted fields provides habitats for predators of crop pests and thus reduces the need for artificial predator control. Retention of forest patches in Costa Rica halved the damage by coffee borer beetles by encouraging bird predation.
- A much-quoted example is the savings by the City of New York of around $7 billion capital cost and $300,000 annual operating costs by restoring and then using the Catskill Watershed for 'natural' water purification.

The MA represented the consensus view of the largest body of social and natural scientists ever assembled to assess knowledge of the natural world. Their focus on ecosystem services and the link to human well-being and development needs was unique. By examining the environment through the framework of ecosystem services, it was much easier to identify how changes in ecosystems may influence human well-being and provide information in a form that decision makers can weigh alongside other social and economic information.

The MA identified a number of 'emergent' findings. Four of these are worth noting:

1. *The balance sheet.* Although individual ecosystem services had been assessed previously, the recognition that 60 per cent of the twenty-four ecosystem services examined by the MA are being degraded provided the first comprehensive audit of the status of the Earth's natural capital.
2. *Nonlinear changes.* The MA found that nonlinear ecosystem changes (that is changes which are significantly faster than average) are increasing the likelihood of damaging effects. This may have important consequences for human well-being, such as, for example, the appearance of new diseases, abrupt alterations in water quality, the creation of 'dead zones' in coastal waters, the collapse of fisheries and shifts in regional climate.
3. *Drylands.* Because the MA focused on the linkages between ecosystems and human well-being, it revealed a new set of priorities. While confirming such well-known concerns as damage to tropical forests and coral reefs, the most significant challenge was found to involve dryland ecosystems. These are particularly fragile, and are also the places where the human population is growing most rapidly, poverty is most acute and biological productivity is lowest.
4. *Nutrient loading.* The MA confirmed the emphasis already being given to important drivers of ecosystem change such as climate change and habitat loss. But it also found that excessive nutrient output from agrochemicals and animal waste is one of the major current drivers of change and will grow significantly worse in the coming decades unless action is taken. Although well studied, the issue of excessive nutrient loading was seen as not yet receiving significant policy attention in many countries.

At the same time that the MA was getting under way in 2000, the world's leaders made a 'Millennium Declaration', spelt out in eight Millennium Development Goals to be achieved by 2015. The seventh Goal was to 'ensure environmental sustainability' and called for (1) new government policies to integrate sustainable development into national policies and programmes, (2) a reduction in the rate of biodiversity loss, (3) better access to safe drinking water and basic sanitation and (4) improving the lives of at least 100 million slum dwellers by 2020.

By identifying targets in this way, the Millennium Development Goals drew attention to the impacts that humans were having on their environment. They were partially successful, particularly in achieving better sanitation and in reducing poverty, but they have now been superseded by a set of seventeen Sustainable Development Goals with 169 detailed targets, which include continuing commitments 'urgently' to combat climate change and its impacts; conserve and sustainably use the oceans, seas and marine resources for sustainable development; protect, restore and promote sustainable use of terrestrial ecosystems; sustainably manage forests; combat desertification; and halt and reverse land degradation and diversity loss.

Such goals are worthy aims. The trouble is that they are rather like yet another attempt at a ceasefire in an otherwise unresolved war. Will they change the situation revealed by the MA? At least they give a yardstick to monitor progress.

- Over the past fifty years humans have changed ecosystems more rapidly and extensively than in any comparable period of time in human history, largely to meet rapidly growing demands for food, fresh water, timber, fibre and fuel. For example, 20 per cent of the world's coral reefs have been lost in the last few decades and another 20 per cent degraded; 35 per cent of the area of mangroves has been lost; withdrawal of water from rivers and lakes has doubled; more than half of all synthetic nitrogen has been applied as fertilizer since 1985.
- Changes made to ecosystems have contributed to substantial net gains in human well-being and economic development, but these gains have to be set against the expense of increased costs from the degrading of many ecosystem services, growing risks of nonlinear changes and the

exacerbation of poverty for some groups of people. Food production has more than doubled since 1960 and its price has dropped, but 25 per cent of commercial fish stocks are being overharvested (the marine fish catch has been declining since the late 1980s).

- The degradation of ecosystem services could grow significantly worse in the coming decades. It has been a barrier to achieving the Millennium Development Goals. More than 15 per cent (and possibly as much as 35 per cent) of water withdrawn for irrigation is exceeding the rate of replenishment. In many areas pest control by natural enemies has been replaced by artificial pesticides, reducing the capacity of natural systems to provide pest control. There is also a global decline in the abundance of pollinators. Once a threshold of nutrient loading is exceeded, changes in freshwater and coastal ecosystems can be both quick and devastating, creating harmful algal blooms (including some produced by toxic species) and sometimes leading to the formation of oxygen-depleted zones, killing all animal life. The loss of wetlands and mangroves and deforestation have reduced the capacity for natural buffering against extreme events. The economic value of a natural ecosystem is almost always higher than a manipulated one (e.g. traditional forest use over timber harvesting; protection to mangroves over shrimp farming), although local gains may muddy the situation (e.g. tourist developments). Deforestation generally leads to decreased rainfall. The existence of healthy forests depends on rainfall, which means that forest loss can result in a positive feedback and exacerbate the situation – accelerating the rate of decline in rainfall, which in turn may lead to a nonlinear change in forest cover. Desertification affects the livelihoods of millions of people, including a large portion of the poor in drylands. The collapse of the Newfoundland cod fishery in 1992 led to an estimated $2 billion cost to provide income support and retraining. The frequency and impacts of floods and fires has increased significantly in the last fifty years, in part at least to environmental changes. Half the urban population in Africa, Asia, Latin America and the Caribbean suffers from one or more diseases associated with inadequate water and sanitation.

The challenge to reverse the degradation of ecosystems at the same time as meeting increasing demands for their services could be met in part under some scenarios suggested by the authors of the

MA, but all of them involve significant changes in policies. A further 10–20 per cent of grassland and forest is likely to be converted to agricultural use by 2050, while overfishing continues virtually everywhere. Invasive alien species are causing increasing problems. At the same time, the demand for food crops is projected to grow by 70–85 per cent by 2050, and water withdrawal by 30–85 per cent.

The MA and the targets in the Millennium Development and Sustainable Goals are part – but only a part – of an increasing international perception that we are living off natural capital, and extracting an unsustainable amount of it. As already noted, this alarm was sounded by the Stockholm Conference in 1972 and reinforced by the World Conservation Strategy in 1980.

BOX 9.5   **Religion and the environment**

Environmental concerns are much more complicated than can be dealt with in a simple cost–benefit analysis. We are told that belief in so-called 'higher gods' (defined as 'supernatural beings believed to govern reality') is more prevalent in societies living under some sort of stress – from climate instability, resource abundance, political complexity.[1] There are clear empirical associations between religious belief and environmental forces. Indeed, the evidence is that a shared belief in a 'moralizing high god' can improve a group's ability to deal with environmental duress. Is religious belief a by-product of stress or are the two causally linked? As always with statistical correlations, this question is difficult to answer. Certainly, religion and environmental concern often occur together. For example, 15 per cent of the world's forests are regarded as sacred; 5 per cent of the world's commercial forests are actually owned by a religious body.[2] The World Wide Fund for Nature (WWF) and the Club of Rome have joined with the Alliance of Religions and Conservation (p. 209) to emphasize the enormous areas of land regarded as a 'sacred space' by different cultures and different religions. Jews, Buddhists, Muslims and Christians have all undertaken initiatives for environmental care within their own traditions, with varying impact on their followers. 'Stewardship' frequently occurs in their statements, although more often honoured in the breach than the reality. For example, the word *khalifah* ('steward') occurs often in the Quran, but 'is curiously underrepresented in the scholastic tradition of Islamic theology'.[3]

Building on the Declaration of Human Rights in a way parallel to the call of the UN Secretary-General, the German theologian Hans Küng proposed a specifically multifaith initiative, emanating from the 'Parliament of the World's Religions' meeting in Chicago in 1993. This proclaimed:

> We are convinced of the fundamental unity of the human family on Earth. We recall the 1948 Universal Declaration of Human Rights of the United Nations. What it formally proclaimed on the level of rights we wish to confirm and deepen from the perspective of an ethic... Earth cannot be changed for the better unless the consciousness of individuals is changed. We pledge to work for such transformation in individual and collective consciousness, for the awakening of our spiritual powers through reflection, meditation, prayer, or positive thinking, for a conversion of the heart. Therefore we commit ourselves to a common global ethic, to better understanding, as well as socially-beneficial, peace-fostering, and Earth-friendly ways of life.

A specifically Christian response came from another German theologian, Jürgen Moltmann. At the 1983 Assembly of the World Council of Churches he argued that the traditional Christian call for 'peace with justice' was futile unless it took place within a *whole* creation, a creation with 'integrity'. His advocacy led to the replacement of the concept of a 'Just, Participatory and Sustainable Society' which had dominated World Council of Churches policy in the 1970s by a more inclusive 'Justice, Peace and the Integrity of Creation' Programme (JPIC). This recognized that environmental concern should be seen as integral to 'justice' and 'peace'; the concept of the 'integrity of creation' was intended to convey the dependence of creation on its creator and the worth and dignity of creation in its own right. It was welcomed by developing countries, who had (rightly or wrongly) associated 'sustainability' with the maintenance of colonial injustice; for them, JPIC implied the rejection of a global hegemony in favour of regional associations.

The JPIC process progressed with different emphases and hopes in different parts of the world. It culminated in a global consultation in Seoul in 1990, which revealed more discord than harmony. A significant proportion of participants saw meaning in creation through mysticism or traditional religions; such theological pluralism was unacceptable to many of those present.

A more inclusive initiative took place when representatives of the main world religions met with the leaders of the international conservation movement in Assisi in 1986 as part of the twenty-fifth anniversary of WWF, and shared their beliefs regarding the moral imperative to care for the environment. These were published as a set of Assisi Declarations; this work has continued as the Alliance of Religions and Conservation.[4] Then, in the early 1990s, a number of independent factors led to the establishment in the USA of a National Partnership for the Environment – these factors were:

- an open letter to the American Religious Community, spearheaded by Carl Sagan and signed by thirty-two Nobel laureates;
- a statement by Pope John Paul II on *The Ecological Crisis, a Common Responsibility*;
- a pastoral statement 'Renewing the Earth' from the US Roman Catholic bishops;
- a consultation on 'environment and Jewish life' in Washington DC in 1992, leading to the formation of the Coalition on the Environment and Jewish Life;
- the publicity and momentum surrounding the 1992 Earth Summit.

Meanwhile, in Europe, in 1989 the Ecumenical Patriarch dedicated the first day of September as a day of protection for the environment. His successor, Bartholomew I of Constantinople, has enthusiastically developed this concept and has hosted a series of international and interdisciplinary symposia to force the pace of religious debate on the environment.

This is not the place to detail all the religious concerns for the environment, but it is worth noting the 2015 encyclical *Laudato Si'* by Pope Francis, arguing that care for the planet was a specifically moral issue. A parallel call from evangelical Christians had come in 2012 as a 'Creation care call to action'.[5]

As noted above (p. 208), the call by the UN Secretary-General for 'a covenant regulating relations between humankind and nature' was answered by the International Environmental Law Commission, with an International Covenant on Environment and Development (ICED), intended to codify into 'hard law' the large amounts of 'soft law' on the environment contained in a range of agreements and treaties such as the Stockholm Conference, the World Charter for Nature, the Law of the Sea and the Rio Declaration. The draft ICED has been through various revisions, most recently in 2004. It still awaits formal adoption by the international community. Notwithstanding, it represents the beginning of a formal convergence between the limited concerns of humankind and the wider concerns of the natural world.

The ICED consists of a preamble and seventy-two articles, arranged in eleven sections. Part I states its object ('To achieve environmental conservation and sustainable development by establishing integrated rights and obligations'), while Part II contains 'the most widely accepted and established concepts and principles of international environmental law'. A significant feature is that the early parts of the ICED contain statements which have repeatedly appeared in the conclusions of other international discussions and agreements – most importantly the Declaration produced at the end of the Stockholm Conference, the World Charter for Nature and the Rio Declaration, but these are only the tip of an iceberg. At least seventy 'statements' or 'agreements' listing an ethic or movements towards an ethic have been produced.[6] Intriguingly, the principles set out in the early parts of ICED are very close to the 'Basic Principles' of the Earth Charter, which was published in 2000.

[1] Botero, C.A., Gardner, B., Kirby, K.R. et al. (2014). The ecology of religious beliefs. *Proceedings of the National Academy of Sciences*, **111**: 16784–9.

[2] Palmer, M. and Finlay, V. (2003). *Faith in Conservation*. Washington, DC: World Bank.

[3] Serageldin, I. (1991). The justly balanced society. In: *Friday Morning Reflections at the World Bank*. Washington, DC: Seven Locks Press, p. 62.

[4] Edwards, J. and Palmer, M. (eds.) (1997). *Holy Ground*. Yelverton: Pilkington Press.

[5] Bell, C. and White, R.S. (eds.) (2016). *Creation Care and the Gospel: Reconsidering the Mission of the Church*. Grand Rapids, MI: Hendrickson.

[6] Bakken, P.W., Engel, G. and Engel, J.R. (1995). *Ecology, Justice and Christian Faith*. Westport, CN and London: Greenwood.

BOX 9.6   **The Earth Charter**

An Earth Charter was one of the original aims of the Earth Summit, intended to produce 'a short, uplifting, inspirational, and timeless expression of a bold new global ethic', containing 'the basic principles for the conduct of nations and peoples with respect to environment and development to ensure the future viability and integrity of the Earth as a hospitable home for human and other forms of life'. It was an aspiration that disappeared in pre-conference wrangling; the Charter was changed into a less focused set of statements, issued and agreed as the 'Rio Declaration'.

This was a disappointment to many, and after Rio, Maurice Strong (who had been Secretary-General of both the Stockholm and Rio Conferences) and ex-Russian President Mikhail Gorbachev both launched organizations to 'promote the transition to sustainable ways of living and a global society founded on a shared ethical framework that includes respect and care for the community of life, ecological integrity, universal human rights, respect for diversity, economic justice, democracy, and a culture of peace'. The two organizations came together in 1994 and a draft Earth Charter was published three years later.

The Earth Charter set out sixteen principles in four groups:

I. **Respect and care for the community of life**

1.   Respect Earth and life in all its diversity.
2.   Care for the community of life with understanding, compassion and love.

3. Build democratic societies that are just, participatory, sustainable and peaceful.

4. Secure Earth's bounty and beauty for present and future generations.

II. Ecological integrity

5. Protect and restore the integrity of Earth's ecological systems, with special concern for biological diversity and the natural processes that sustain life.

6. Prevent harm as the best method of environmental protection and, when knowledge is limited, apply a precautionary approach.

7. Adopt patterns of production, consumption and reproduction that safeguard Earth's regenerative capacities, human rights and community well-being.

8. Advance the study of ecological sustainability and promote the open exchange and wide application of the knowledge required.

III. Social and economic justice

9. Make the eradication of poverty an ethical, social and environmental imperative.

10. Ensure that economic activities and institutions at all levels promote human development in an equitable and sustainable manner.

11. Identify gender equality and equity as prerequisites to sustainable development and ensure universal access to education, health care and economic opportunity.

12. Uphold the right of all, without discrimination, to a natural and social environment supportive of human dignity, bodily health and spiritual well-being, with special attention to the rights of indigenous peoples and minorities.

IV. Democracy, nonviolence and peace

13. Strengthen democratic institutions at all levels, and provide transparency and accountability in governance, inclusive participation in decision making and access to justice.

14. Integrate into formal education and life-long learning the knowledge, values and skills needed for a sustainable way of life.

15. Treat all living beings with respect and consideration.

16. Promote a culture of tolerance, nonviolence and peace.

The Basic Principles of the ICED can be reduced to ten premises.[8] The commonality between these independent sources and their later expression in the Earth Charter suggest that they may well represent truly fundamental principles rather than convenient or arbitrary generalization.

BOX 9.7　**Ten premises for sustainable living**

In the following the numbers refer to paragraphs in documents: 'S' refers to the Stockholm Declaration of the UN Conference on the Human Environment in 1972; 'C' to the World Charter for Nature agreed by the UN in 1982; and 'R' to the Rio Declaration accepted by the Earth Summit in 1992.

1.　Environmental conservation and sustainable development are essential for human health and well-being on a planet with finite resources and carrying capacity (S13; C2; R1, 4).

2.　Nature as a whole warrants respect; every form of life is unique and is to be safeguarded independently of its worth to humanity (C Preamble, 1).

3.　The global environment both within and beyond the limits of national jurisdictions is a common concern of humanity, held in trust for future generations by the present generation. All persons have a duty to protect and conserve the environment; each generation has a responsibility to recognize limits to its freedom of action and to act with appropriate restraint, so that future generations inherit a world that meets their needs (S2, 4; C3; R4).

---

[8] On a personal note: I was a member of a small group of scientists from the Ethics Working Group of IUCN which met with the Environmental Law Commission, who had recognized the need for scientific scrutiny of their work on the ICED. I urged that the 'Basic Principles' in the draft be treated as a separate 'treaty', which would be more easily endorsed by governments than a detailed 'charter' with specific prescriptions on limits, fines, etc. Acceptance of the charter could follow more easily after the principles were accepted. This proposal was rejected by the lawyers who wanted to maintain the ICED as a single document. I discussed the contents of the proposed treaty with Steven Rockefeller (also a member of the Ethics Working Group). He went on to be the lead agent in drafting the Earth Charter, which sets out the same ideas and balance as in the 'Basic Principles' derived from the ICED. This convergence of concepts is a strong ground for believing that they represent a description of underlying reality, not simply the product of argument.

4. To achieve sustainable development, environmental protection and management must be an integral part of all development efforts. States have, in accordance with the Charter of the United Nations and the principles of international law, not only the sovereign right to their own resources, but the responsibility:

    a. to protect and preserve the environment within their jurisdiction or control;

    b. to ensure that activities within their jurisdiction or control do not cause serious damage to the environment of other states or to areas beyond the limits of national jurisdiction;

    c. to work with and collaborate in good faith with other states and competent governmental and non-governmental organizations in the implement of the ICED;

    d. to minimize waste in the use of natural resources and ensure that renewable natural resources are maintained sustainably, and to develop and adopt the most efficient and environmentally safe technologies for the harnessing and use of energy (S7, 14, 21; C4, 7; R4, 12).

5. All states and all people shall cooperate in promoting health, social well-being and environmental quality by striving to eradicate poverty; this is an indispensable requirement for both sustainable development and distributive justice, and can be achieved only by eliminating unsustainable patterns of production and consumption and by promoting appropriate demographic policies (R5, 7, 8).

6. States have a responsibility to anticipate, prevent and minimize significant adverse effects of human activities on the environment; lack of full scientific certainty must not be used as a reason to postpone action to avoid potential harm to the environment (C11; R15, 18, 19).

7. States shall take all necessary measures to ensure that the full costs of prevention or compensation for environmental damage, as well as the costs of restoration of the environment, are borne by the person or organization whose activities give rise to such damage or the threat thereof, unless such obligations are otherwise allocated by national or international law. States have the right to be protected against or compensated for significant environmental harm caused by activities outside their own jurisdiction (S22; C12; R13).

8. States shall require environmental impact assessments for all proposed activities likely to have a significant environmental effect and shall

include the full social and environmental costs of all environmental impacts within the calculation of those effects (C11; R17).

9. States shall establish and maintain a legal, administrative, research and monitoring framework for environmental conservation, giving full and equal consideration to environmental, economic, social and cultural factors. In particular, states shall:

   a. regularly review their policies on the integration of planning and development activities and publish their findings;

   b. develop or improve mechanisms to facilitate the involvement of concerned individuals, groups, organizations, indigenous peoples and local communities in environmental decision making at all levels, and provide effective access to judicial and administrative proceedings affecting the environment;

   c. make clear the full social and economic costs of using natural resources and ensure the equitable distribution of income generated (S18, 20; R9, 10, 22).

10. Justice, peace, development and environmental protection and management are interdependent and indivisible, and vital to the integrity of creation (S1; R25). States have a responsibility to work towards an environmentally aware citizenry that has the knowledge, skills and moral values to protect and preserve the environment and to achieve sustainable development.

The first two premises bring together as essential complements human responsibility with what is best described as awe and wonder, responses which can be more easily described in poetic or religious language than in the words of scientists or lawyers. They focus on an attitude to nature which is neither a claim to 'rights' (whether human or nature's) nor an assertion of dominion. Premises 3 and 4 are also complementary, asserting the role of individuals and states respectively. In addition, premise 3 recognizes the existence and importance of the 'global commons' and of 'transgenerational equity' (see Chapter 2, pp. 20–46); premise 4 lays an additional charge on states to develop renewable energy.

Premise 5 is a commitment to eradicate poverty. This is a key condition for developing nations, but this premise embeds it firmly in the often ignored context of sustainability and demography. Economists tend to express this principle in terms of maintaining capital, regarding it legitimate to substitute a declining asset by one more readily available (such as plastic for metal, wind or wave energy for fossil fuel). Premises 6, 7 and 8 are three widely accepted and almost as widely ignored principles: the precautionary principle, the polluter pays principle and environmental impact assessment. The last of these is a necessary component of the precautionary principle, but it warrants inclusion, not least because it implies monitoring (premise 9), which is especially important in view of the slow and often insidious nature of environmental change. Premise 10 returns to the necessary partnership between state and its citizens, which is an essential part of any functional code of conduct.

The ten premises bring us firmly back to the notion of stewardship. They describe responsibilities rather than rights. There are both practical and theoretical reasons for this: responsibility involves response from those able to influence actions, whereas a right is merely a status; responsibility is implicit in the relationship between states, people, other living beings and the Earth and its environment as a whole. Rights language implies a static relationship between rights-givers and rights-claimers, whereas the relationship between humankind and environment is a dynamic one. Rights ought to be anathema to any environmentalist concerned with exercising responsible care and protection of the natural world. Moreover, they are prejudicial to developing a common approach, because they establish categories for protection rather than concentrating on the processes which determine the categories.

BOX 9.8   **Does nature have 'rights'?**

'Rights language' in general is most commonly used to express dissent from political or social restriction by (or on behalf of) those who are oppressed or underprivileged. Some see that adherence to legally enshrined rights is essential for a fairer, more secure society – a fundamentalist belief to be defended at all costs. It is a belief stemming from Immanuel Kant and the assumptions of the French (and then the American) Constitution. It became unfortunately complicated by references to the basic Four Freedoms – of speech, religion, fear and want, codified into the 'Universal Declaration of Human Rights' (the most translated document in the world, according to the Guinness Book of Records) adopted by the UN in 1948 to be promoted 'without distinction as to race, sex, language, or religion'. This was seen as a Bill of Rights for the human race; it has been described as a 'Magna Carta for Mankind'. Without dissenting one jot from the importance of these freedoms, it must be noted that they are really nothing more than a legal template, and certainly not universal. They represent aspiration rather than substance, an aim for a Utopia. The Declaration does not explain where human rights comes from; it takes them as self-evident, just as did Jefferson's Declaration of the Rights of Man at the end of the eighteenth century. At best they can be regarded as being derived from a notion of individual freedom and responsible relationships common to ancient civilizations in both east and west, part of the common good recognized by Plato, Aristotle and Cicero (p. 2). But this forces us away from legal or moral questions to ask 'what does it mean to be human?' – which is where we began.

Another problem is that in practice rights tend to be asserted in a much cruder sense – in an adversarial and assumedly non-negotiable way. This makes them a lawyer's delight. Debates about rights have had the bonus of opening philosophy to environmental issues, but they detract from the responsibilities we have for our actions and the privileges which flow from them. They should sound a warning to anyone concerned with responsible care and protection for the natural world. Furthermore, they erect barriers by establishing categories for protection rather than concentrating on the processes that determine the categories.

Notwithstanding, 'rights' are frequently asserted in discussions about the environment. Their earliest use in environmental matters seems to have been by John Muir, who wrote: 'How narrow we selfish, conceited creatures are in our sympathies! How blind to the *rights* of all the rest of creation!' In an address to the UN General Assembly in 2015 and in his associated encyclical, Pope Frances was emphatic:

> It must be stated that a true 'right of the environment' does exist, for two reasons. First, because we human beings are part of the environment. We live in communion with it, since the environment itself entails ethical limits which human activity must acknowledge and respect... Any harm done to the environment, therefore, is harm done to humanity. Second, because every creature, particularly a living creature, has an intrinsic value, in its existence, its life, its beauty and its interdependence with other creatures. We Christians, together with the other monotheistic religions, believe that the universe is the fruit of a loving decision by the Creator, who permits man respectfully to use creation for the good of his fellow men and for the glory of the Creator; he is not authorized to abuse it, much less to destroy it. In all religions, the environment is a fundamental good.[1]

The modern concept of environmental rights came to the fore in the 1970s following the granting of permission by the US Forestry Service to Walt Disney Enterprises for a leisure complex in the Californian Sierra Nevada. The Sierra Club appealed against this, but lost on the grounds that it was not a plaintiff in the case and therefore had no legal right to be heard. A Californian lawyer, Christopher Stone, questioned this. He asked 'Should trees have standing?' (*South Californian Law Review*, **45**: 450–501, 1972). His reasoning was that something was in danger of injury by the development, and that the courts should be sensitive to the need for protection. He believed that society should 'give legal rights to forests, oceans, rivers and other so-called "natural objects"' – indeed, to the natural environment as a whole. He recognized that trees and rivers could not institute proceedings on their own behalf, but suggested that extending the idea of trusteeship could give them 'standing' in the legal sense. He contended that natural objects had definite needs, the denial of which resulted in perceptible deterioration. Stone suggested that the damage from (for example) polluted air or water could be quantified and recompense collected by human guardians on behalf of the affected air or water. One of the judges in the appeal against planning consent commented, 'Contemporary public concern for protecting nature's ecological equilibrium should lead to the conferral of standing upon environmental objects to sue for their own protection.'

The concept of environmental rights remains contentious. The philosopher Mark Sagoff has firmly rejected the notion but suggested a way forward is to accept a 'weak anthropocentric morality', that there is 'a right to wild mountain valleys for their cultural, spiritual and aesthetic value – to people'.

[1] Encyclical *Laudato Si'* (2015). English translation published by London: Catholic Truth Society, para. 81.

It would be foolish to think that environmental care will be achieved solely by ethical pronouncements or declarations from international meetings, but it does seem that some universal guiding values have emerged over the last few decades. Furthermore, these values link what strikes the mind and heart as sound principles with the fear of what may happen if we ignore them. Such universal themes are a recognition:

- that long-term sustainability must not be avoided; it is an essential object for policy now;
- that equity between people and nations on their impact on the finite resources and vulnerable resources of the planet must be improved; and
- that we now have a way to codify and communicate our obligations to other people and the world as a whole.

Notwithstanding, the likelihood that there are indeed 'basic principles' raises a hope that there might be a way forward. It seems clear that they are not an arbitrary set of statements; they descend from and depend on earlier gropings towards a global ethic, from at least the time of John Ruskin and John Muir, and extend through much international debate. Beginning from the awareness that we are running out of world and its quantification in the computer simulations of the *Limits to Growth* studies, the concerns about poverty highlighted at the Stockholm Conference and followed by the ideals expressed in the World Charter for Nature agreed by the UN in 1982, there has been a progression to *Caring for the Earth* of 1991, with its clear demonstration that sustainable development is impossible without reliable environmental care. Nevertheless, we have to accept that, while this logical connection is necessary, it is not sufficient. It needs personal and political commitment. Effective environmental care is only attainable through responsible management on the part of both individuals and organizations at all levels. We may be apart from nature, but we are also a part of it.

FURTHER READING

Bourdeau, P., Fasella, P.M. and Teller, A. (eds.) (1990). *Environmental Ethics: Man's Relationship with Nature, Interactions with Science.* Luxembourg: Commission of the European Communities.

Cairncross, F. (1991). *Costing the Earth.* London: Business Books.

Carson, R. (1962). *Silent Spring.* Boston, MA: Houghton Mifflin.

Grubb, M., Koch, M., Munson, A., Sullivan, F. and Thomson, K. (1993). *The Earth Summit Agreements.* London: Earthscan.

Helm, D. (2015). *Natural Capital: Valuing the Planet.* New Haven, CT and London: Yale University Press.

Holdgate, M.W. (1996). *From Care to Action.* London: Earthscan.

Holdgate, M.W. (1999). *The Green Web.* London: Earthscan.

Juniper, T. (2015). *What Nature Does for Britain.* London: Profile Books.

Lear, L. (1997). *Rachel Carson: Witness of Nature.* New York: Henry Holt.

Sagoff, M. (1988). *The Economy of the Earth.* Cambridge: Cambridge University Press.

10 Reckoning, Perhaps Rueing

Where does all this get us? Does knowing about past arguments or faraway agreements help to deal with the present, to form and strengthen a proper attitude to help *my* decision making? To anticipate our destination, we seem to be coming full circle and approaching the place where we started: a need to recognize and then take responsibility for our own actions. Over the centuries this responsibility has moved from our personal sphere to governments and on to international agencies. We can obviously shrug off any claims the environment has on us, but in so doing we may well be mortgaging the future of ourselves and our children. We have to face moral questions.

Are there right and wrong attitudes to the way we view the environment? This is not a hypothetical question but one that needs an answer if we are to be prepared for the 'perfect storm' which seems to be approaching (p. 22). It is only too easy to regard documents like the World Conservation Strategies or measures to deal with climate change as worthy but irrelevant to 'normal' people. Is this true? One way of looking at this is to revisit the response to the original World Conservation Strategy when it appeared in 1980. One of its sponsors was UNEP, and there was a requirement for all member nations of UNEP to react and draw up their own national strategies for conservation. In the UK this task was the responsibility of the (then) Nature Conservancy Council. The initial idea was to welcome the Strategy with as much pomp and publicity as possible. Hiring the Albert Hall, the Duke of Edinburgh and a Brigade of Guards Band were mooted. A more mature and positive second thought was to enquire how different parts of the Strategy applied in Britain. Working parties were established to examine their relevance and possible impact.

The reports of the seven working parties were published together in 1983 as the formal UK Response, dealing with the topics raised in the Strategy itself: industry, urban, rural, marine and coastal, international, education and ethics.[1] A novelty was the seventh section, dealing with ethics. As already noted (p. 195), the Strategy itself assumed that knowledge of needs would automatically elicit a response; it omitted any discussion of ethics. Max Nicholson decreed there should be an ethics section in the overall report and spelt out the need in his foreword to the UK Response:

> Well publicised Doomsday statements [have] launched a strong challenge to the myths of unending exponential economic growth and shocked the public into a recognition of the hitherto ignored limits of the Earth... It may be claimed that no other country has responded to the World Conservation Strategy in so much depth and breadth... Civilisations which persist in pressing too single-mindedly for too long on a line of advance, however broad, end up in a blind alley. That happened, for example, to medieval Christendom just before the Renaissance, and it is happening to our lopsided technological modernism now, both in East and West. Human evolution builds up seismic tensions at great depth, which can only be resolved by eventual creative tension between apparently irreconcilable sets of values.

The ethics section of the UK Response was prepared by a group chaired by (Lord) Eric Ashby, a botanist and university administrator, who became the first Chairman of the UK's standing Royal Commission on Environmental Pollution, set up in 1970 by Prime Minister Harold Wilson in the wake of European Conservation Year and as part of the government's preparations for the Stockholm Conference.[2] In his introduction, Ashby commented that:

---

[1] *The Conservation and Development Programme for the UK. A Response to the World Conservation Strategy*. London: Kogan Page, 1983.

[2] Over its forty-year existence the Royal Commission produced a series of reports, some of which challenged accepted assumptions or administrative inertia (e.g. on nuclear power, water quality, agricultural pollution, lead in the environment,

while [the Strategy as a whole] benefits (or at least does not inconvenience) the UK, its citizens will welcome the recommendations
[in the report as a whole] to upgrade industry's environmental
resources, to control the disposal of toxic wastes, to reduce SO_2
[sulphur dioxide] emissions from power plants, to give aid in developing countries, to propagate environmental education in schools,
and to conserve our rural resources... but at a time of stress and
austerity these issues are likely to become the first casualties...
Until we have a strategy for managing the conflict between the
homocentric and ecocentric aspirations of humanity, our good
resolutions about the survival of the earth will remain precarious.

The ethics report was based on a simple premise: that we are
simultaneously apart from nature and a part of nature. This was not
original; it is a theme which has recurred throughout this book (it
surfaced in the Pope's encyclical *Laudato Si'*): from our origins as an
evolving ape able to do no more than respond to its environment to
a being capable of largely insulating itself from its environment. But
this process of emancipation is an implicit recognition of decisions
putting some sort of quantitative value on the environment. This is
anathema to many. What does it mean in practice? Even at the individual level, the environment has value in at least four ways which
reflect and support the elements which add up to well-being (p. 198):

- cost in the market-place, quantifiable as money – farming, forestry,
  mining, proprietorship;
- usefulness – by providing water, food, fuel and 'services' (recycling,
  cleaning air and water, pest control, crop pollination);

etc.). A major, albeit unstated, function of the reports was to produce a political
accommodation with science, society and regulation. Importantly, they could
not be ignored: each required a government response. There is no doubt that the
Commission influenced environmental attitudes through the reports themselves
or through subsequent government action. The Commission was abolished by
its political masters in 2011, on the grounds that they could get environmental
advice elsewhere (Owens, S. (2012). Experts and the environment – the UK Royal
Commission on Environmental Pollution 1970–2011. *Journal of Environmental
Law*, **24**: 1–22).

- intrinsic worth, depending on the objective quality of the object valued, in contrast to the market-place cost which is measured only in relation to price of other things that can be acquired in its place;
- symbolic or conceptual worth – as with a national flag or freedom itself.

A further complication is that the same physical entity can carry all these values. For example, a plot of land has a market value; it has value-as-use for a farmer or home-owner; it may have intrinsic value for its beauty; and it may be valuable as a symbol of territory to be defended against enemies. An even further difficulty is that these four meanings may change independently for the same object: a stream in a rural highland will be valued differently from one in an urban lowland according to its different uses – whether it is drunk, fished or treated as an amenity; whether it is an object of beauty or a stinking sewer; whether it is a boundary between estates or a barrier to pest spread; and so on.

What are the stresses spoken of by Ashby and how do they relate to environmental values? They are essentially the assaults all of us have faced from our pre-human days: the retreating forest which forced our forebears to become plain dwellers, the changing climate which periodically encouraged us to migrate or forced us to retreat, the crises of floods or disease, conflicts with predatory animals or poachers or greedy neighbours. These are the stresses described by Jared Diamond in *Collapse*, which have contributed to the decline and sometimes disappearance of past cultures (p. 185). Examples include overpopulation, overextended irrigation, overharvesting of local resources and changing weather patterns. Sometimes the stress is immediate and catastrophic (earthquake, volcanic eruption, epidemic disease) but more often and more dangerous is when the stress is chronic and cumulative, particularly if we resist it for greed – personal or institutional. The moas on which the Maoris depended did not suddenly disappear but must have become harder and harder to find. Yields from both coastal waters and the deep sea have dropped depressingly as fish populations have been depleted.

Increased effort and ever-improving technology have compensated for this, but there comes a time when the return is so small as to be non-economic. This happened with commercial whaling – first in the North Atlantic and then in the Antarctic. On top of all this is the reluctance of many to accept the reality of climate change, or at least any human responsibility for it.

BOX 10.1   **Ecosystems and equilibria**

Through his emphasis on natural law, Thomas Aquinas extended Aristotle's belief in the rationality and purposiveness of the world. Basing himself on the Stoic idea of the creator's wisdom and benevolence, Aquinas taught that there was a balance and harmony in the natural world. This was supported by the commonly applied parallel between the microcosm of the body and the macrocosm of the living world. For Aquinas, religion did not so much as encroach on ecology as envelop it. His successors accepted this without question; they took it for granted that communities of animals and plants existed as natural, repeated, internally organized units with a considerable degree of integration. Such a community was commonly regarded as a super-organism, or at least quasi-organismic. Early ecological scientists did not challenge this notion. The US botanist Frederic Clements (1874–1945) was wont to argue that, since there was an inevitable development in plant communities towards a climax community, this could be regarded as a super-organism. His ideas were developed philosophically by Edinburgh-born botanist John Phillips, who sought to show that 'in accordance with the holistic concept [of South African Jan Christian Smuts, 1870–1950], the biotic community is something more than the mere sum of its parts; it possesses a special identity'.[1] Smuts had devoted two years to writing *Holism and Evolution*, in which he stated his belief that organisms represent microcosms in which member cells cooperate in a regime of law and order, and which serve as models for human society. For Phillips, 'the biotic community… is a mass-entity with a destiny peculiar to itself'.

All this was too much for the leading British ecologist of the period, Arthur Tansley (1871–1955). He argued that the succession from bare soil to mature woodland depends on the plants themselves. He firmly rejected the idea of 'holism' as a fundamental concept (he called it a 'mysterious' factor), and with it the assumptions of Clements, Smuts and Phillips. His contribution was to coin a new word, 'ecosystem', defined as:

> the whole system (in the sense of physics) including not only the organism-complex, but also the whole complex of physical factors forming what we call the environment of the biome – the habitat factors in the widest sense. It is the systems so formed which, from the point of view of the ecologist, are the basic units of nature on the face of the earth.[2]

Thus, the plant community should not properly be understood in isolation from its environment.

Sadly, ecosystems – intended by Tansley as no more than a descriptive generality – have spawned a whole sub-discipline with ascribed properties of resilience, persistence, resistance and variability. They are said to have health and needs and to suffer damage – designations more properly applied to organisms. A better although clumsy description is that ecosystems are 'self-organizing systems in which random disturbance and colonisation events create a heterogeneous landscape of diverse species, which then become knitted together through nutrient fluxes and other forms of interaction... some simply having to do with chance and geography.'[3]

Charles Elton (1900–91), one of the founders of animal ecology, was intrigued by the periodic fluctuations of fur-bearing animals, shown by the returns of fur trappers from the Hudson's Bay Company. It was believed that these showed some sort of harmony – a balance of nature – in natural communities. Elton studied these in depth. His answer?

> The 'balance of nature' does not exist, and perhaps never has existed. The number of wild animals are constantly varying to a greater or lesser extent. And the variations are usually irregular in period and always irregular in amplitude... [The] key to community structure may lie in

> extrinsic biogeography rather than in intrinsic local processes, making community ecology a more historical science.[4]

In other words, there is no such thing as ecological equilibrium.

[1] The biotic community. *Journal of Ecology*, **19**(1): 1–24, 1931.
[2] Tansley, A.G. (1935). The use and abuse of vegetational concepts and terms. *Ecology*, **16**: 284–307.
[3] Levine, S.A. (1999). *Fragile Dominion*. Reading, MA: Perseus.
[4] Elton, C.S. (1930). *Animal Ecology and Evolution*. Oxford: Oxford University Press.

Both individuals and communities can – and usually do – cope with stress, either by mitigation or adaptation, although their responses may be delayed. The leading British ecologist (Sir) Richard Southwood showed how an animal's survival can be reduced to five factors:

- investment to counter inclement physical conditions;
- avoiding death by predation or, for plants, reduced fitness by herbivory;
- food harvesting and development of particular traits for this – such as storage organs in plants;
- reproductive activities – finding a mate, successful breeding, care of offspring (is it better to produce a few large offspring as do elephants or many small ones as do herrings or aphids?); and
- escape – which might be migration or dormancy – to breed in one place or elsewhere (Figure 10.1).[3]

Human strategies come down to the same five basic elements, albeit dressed up in less brutal language.

These interactions produce an apparent pattern in nature which has spawned a whole dictionary of descriptive concepts: mutualism, parasitism, competition, symbiosis, commensalism, antagonism, *ad nauseam*. In fact, Charles Darwin gave a vivid summary of them in his original description of his 'big idea' in 1844:

> Nature may be compared to a surface on which rest ten thousand sharp wedges touching each other and driven inwards by incessant

[3] Southwood, T.R.E. (1988). Tactics, strategies and templets. *Oikos*, **52**: 3–18.

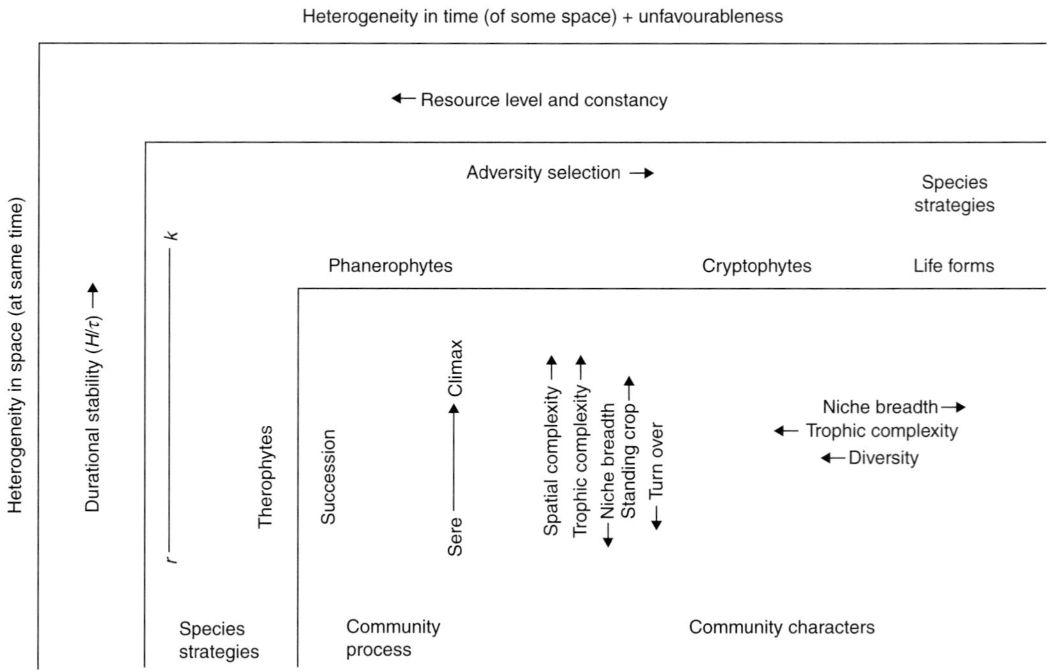

FIGURE 10.1A Relation between habitat type and biological response. Reproduced with permission from Richard Southwood (1977). Habitat, the templet for ecological strategies? *Journal of Animal Ecology*, **46**, 336–65.

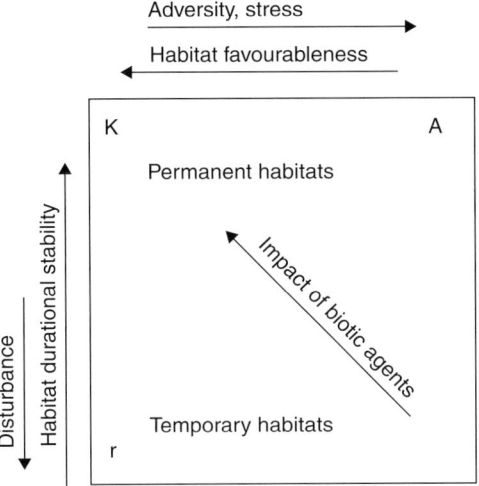

FIGURE 10.1B Different tactics in response to different environmental pressures.
Reproduced with permission from Richard Southwood (1988). Tactics, strategies and templets. *Oikos*, **52**(1), 3–18.

blows. Fully to realise these views much reflection is requisite… But let the external conditions of a country change… can it be doubted from the struggle each individual (or its parents) has to obtain subsistence that any minute variation in structure, habits, or instincts, adapting that individual better to the new conditions, would tell upon its vigour and health? In the struggle it would have a better chance of surviving, and those of its offspring which inherited the variation, let it be ever so slight, would have a better chance to survive.[4]

The examples given by Jared Diamond in *Collapse* show that coping with stress is rarely easy, even when reason shouts the need (p. 185). The nineteenth century was driven by changes in attitude to a Pandora's box of environmental interactions, many of them produced by charismatic individuals linked to public campaigns against a range of social and environmental injustices and insecurities. William Wilberforce was an iconic example in this respect. He

---

[4] Darwin, F. (ed.) (1909). *The foundations of The origin of species. Two essays written in 1842 and 1844*. Cambridge: Cambridge University Press, pp 90–91.

worked for prison reform, better conditions for chimney sweeps and restrictions on capital punishment. He campaigned for education for the poor; he collaborated with Thomas Clarkson for the abolition of slavery. His younger contemporary, the Earl of Shaftesbury, fought for twelve years to persuade parliament to pass a Ten Hours Bill, limiting the employment in mines or factories of those under eighteen to ten hours a day. Opposition to change was sometimes violent, such as the demonstration in Manchester of textile workers against the Corn Law introduced to protect domestic production but thereby increasing food costs, which led to the Peterloo Massacre in 1819. Reform to parliamentary representation led to periodic confrontations and (in retrospect) a painfully slow spread of suffrage and accountable authority.

We have already seen the inaction and prevarication of the British Government about smoke control and the saga of the Plimsoll Line. Like William Wilberforce, Samuel Plimsoll was opposed by a small group of reactionaries with vested interests, who succeeded in compromising marine safety (p. 118). A much more dangerous debate, because it has global, far-reaching and largely irreversible implications, is the impact of climate change. Apart from a very few mavericks, every informed scientist fully accepts that climate change is taking place and that the overwhelming cause of this is the release of massive amounts of CO_2 into the atmosphere from the burning of fossil fuels. There are legitimate debates about the rate and regional effects of climate change, but no reasonable doubts about either its occurrence or dangers. The tragedy is that debate on the subject has been muddled by vested interests, in particular from the oil and coal extraction industries, supported by so-called libertarians – those opposed to any form of regulation. All such denialist groups have been well funded and litigious. This has hindered the assessment of the facts and their interpretation, and made agreement on action considerably more difficult. The issue with such events is separating basic underlying morality from short-term political manoeuvring. How does one decide?

Rational analysis is needed in every such situation, but it is insufficient in itself. There are limits to science.

BOX IO.2 **Reductionism**

That there is a limit upon science is made very likely by the existence of questions that science cannot answer and that no conceivable advances of science would empower it to answer... I have in mind such questions as 'How did everything begin?' 'What are we all here for?' 'What is the point of living?'.[1]

The practice of science involves reductionism, although there is more than one sort of reductionism. The operational reductionism used in examining scientific mechanisms is utterly different from the doctrinaire assumption that knowledge of a mechanism completes all the knowledge about an event.[2] This distinction is not new, although it is frequently ignored (most brazenly by the so-called 'new atheists'). It was spelt out by Aristotle in the fourth century BC when he distinguished four different sorts of causes (material, formal, efficient, final). We rarely separate out all four, but we are very familiar with ultimate and proximate causes (wanting a cup of coffee is an ultimate cause; the heat that warms the water for the coffee is a proximate cause). It is legitimate to concentrate on a particular cause for the purpose of analysis or experiment; it is unhelpful and usually confusing to deny that other causes may be operating. Doctrinaire reductionism (denying that there is more than one cause to an event) is unnecessary and often confusing. The dangers of this are particularly important when we are dealing with human choices. Having to heat water to make a hot drink is necessary, but this does not explain why I wanted a drink in the first place. I may know the full molecular structure of the HIV or Ebola viruses, but that does not prevent or protect me from suffering from them. Too often we choose wrongly because we do not know or distrust the link between cause and effect:

- From ancient times, scurvy killed more seamen than war or shipwreck. It was attributed to many causes. Captain Scott's colleagues suffered from scurvy on his Antarctic expeditions; he accepted the then current belief that it was caused by 'tainted meat', although by the 1860s the juice of citrus fruits was known to prevent it. Indeed, the 'antiscorbutic factor' found in fresh vegetables (which we now call vitamin C) was identified in 1907 before Scott left for the Antarctic.

- The idea of continental drift was originally proposed formally by Alfred Wegener in 1912, but it was widely doubted until the discovery of plate tectonics in 1958 provided a mechanism for its operation.
- Richard Doll showed in 1950 that 'the risk of developing lung cancer increases in proportion to the amount smoked. It may be fifty times as great among those who smoke twenty-five or more cigarettes a day as among non-smokers.' It took until 1964 for the government to ban cigarette advertisements on television and another forty years to ban smoking in public workplaces.

[1] Medawar, P. (1984). *The Limits of Science*. New York: Harper & Row, p. 66.
[2] Ayala, F.J. (1974). Introduction. In: Ayala, F.J. and Dobzhansky, T. (eds.). *Studies in the Philosophy of Biology*. London: Macmillan, pp. vii–xvi.

Conscious environmental understanding has increased enormously in the last few decades, encouraged and driven by events like Earth Day and European Conservation Year; by outstanding communicators like Rachel Carson, David Attenborough and Pope Francis; by international efforts (the Earth Summit, MA, Millennium and now Sustainable Development Goals) and national and international NGOs; and by disasters (earthquakes, hurricanes, tsunamis, droughts and oil spills [*Torrey Canyon, Esso Valdez*, Deepwater Horizon, etc.]), but still there is prevarication and inaction at both political and personal levels. Much has been achieved by national legislation (Environmental Protection Agency and Endangered Species Act in the USA; Wildlife and Countryside Acts in the UK) and international agreements (such as the Habitat, Bird and Water Framework Directives of the EU, the Ramsar Convention on Wetlands, the Convention on International Trade in Endangered Species, the Montreal and London Protocols to control CFCs), by the establishment of protected areas (national parks, nature reserves, exclusion zones for fishing) and by the work of IUCN in monitoring and publicizing biodiversity loss.

Despite all this endeavour, however, the extent of environmental damage – whether expressed in terms of lost natural capital, missed targets on protecting species or habitats, use of dangerous substances (persistent pesticides, plastics which do not degrade) or practices (sea-floor damage through fishing, soil compaction through the use of heavy machinery), greed (shoddy or mis-sited building, unregulated logging, overfishing or grazing) or any other way – is not encouraging. It does not take much insight to recognize that the barriers to responding to ecological changes are not technical; they are almost entirely moral or social (or political). They include:

- inappropriate institutional and governance arrangements, including the presence of corruption and weak systems of regulation and accountability, highlighted by the toleration of flammable cladding of buildings;
- market failures and improper economic incentives;
- social and behavioural factors, not least the lack of political and economic power of many of the groups that are particularly dependent on ecosystem services or are harmed by their degradation;
- underinvestment in developing but more usually the failure of uptake of appropriate technologies;
- insufficient knowledge (as well as the poor use of existing knowledge) concerning ecosystem services and responses that could enhance benefits from these services while conserving resources;
- weak human and institutional capacity related to the assessment and management of ecosystem services.

It is not enough merely to know that human activity can harm the environment. Knowledge must lead to action; however enlightened and rational understanding is, it is useless unless linked to moral (and political) commitment. The utilitarian argument that human flourishing depends on cooperation is true, but is insufficient in itself. For action, we have to accept that we are a part of nature, while at the same time we are apart from nature. This implies what is often called 'holistic thinking', although that term tends to be associated more with being a part of nature rather than apart from it. Perhaps a better way is to go with Peter Medawar (p. 12) and accept the reality and necessity of facing metaphysical questions.

The WWF's Assisi Declarations of 1986 called upon the world's main religions to declare how their teachings led them to care for nature (p. 209). It may be that WWF expected a single powerful statement; what it received was six individual declarations from different faiths. In his presidential address at Assisi, the Duke of Edinburgh distinguished between the practice and motives for environmental conservation:

> It is not enough just to be concerned about the conservation of nature, neither is it enough to have the scientific expertise to enable us to achieve the conservation of nature: we also need a clear and sufficient motive to ensure that our hearts as well as our minds are committed to the cause. We need the knowledge plus commitment. We need a credible philosophy. What we need is to establish the practical and moral reasons why conservation is important, and to clarify the motives that will help people to commit themselves to the cause of conservation... the economic argument... the scientific argument... the moral argument, the relationship between man and nature. There can be little advantage in attempting to save our souls or to seek enlightenment or salvation if our very existence on this earth is threatened by our own destructive activities.

Our existence and ultimate survival as a species is inseparable from our interactions with the world around us. When we first emerged as a species struggling against predators and competitors, against weather and disease, and with the need to find adequate food and shelter, these interactions were not options, they were obligatory for survival. As time passed, we gained options. The villages into which our Neolithic forebears settled gave the possibilities of becoming a storekeeper or builder or perhaps even a sanitary specialist, skilled in drainage. Our development as a species changed from the purely biological to the psychosocial;[5] our attitudes to the environment changed increasingly from involuntary to voluntary. The problem – and tragedy – is that our opportunities have not been accompanied by a maturing in the exercising of our options. We have failed to recognize that choice brings responsibility. My appetites, effluents and habits afflict others. As the Duke recognized, our relationship between man and nature, economically and individually, never mind our communities and children, needs a moral underpinning. It is the same call that Pope Francis made in *Laudato Si'* (p. 209).

The logical case for accepting a responsible environmental attitude seems overwhelming: there is indisputable evidence that we are using up the natural capital of a finite planet and damaging our surroundings through misuse, perhaps irretrievably. This has been spelt out at the institutional level by a succession of international commitments: Stockholm, Brundtland, the World Conservation Strategies, the Earth Summit in Rio, the Climate Change and Biodiversity Conventions, the Earth Charter, the Sustainable Development Goals, Millennium Biodiversity Assessments and so on. They have been supported by a host of more specific agreements – on sustainable harvesting and fishing, protected areas (ranging from national parks and marine exclusion zones to local nature reserves), planning restrictions, species protection, water extraction orders, etc. All these commitments are commonly opposed by calls recognizing the need for economic development for coal mining, timber or oil extraction (including fracking), housing, tourist developments, airport expansion, agricultural expansion and pesticide use, never mind illegal fishing or deforestation, or wilful dumping of pollutants. Environmental protection is attacked as self-serving and anti-economic, or a luxury for the well off. Or impracticable. But the words of G.M. Trevelyan still resound, although they were written almost a century ago: 'It is no less essential for any national health scheme to preserve for the nation walking grounds and regions where

---

[5] Julian Huxley's term (see *Evolution in Action*. London: Chatto & Windus, 1953).

young and old can enjoy the sight of unspoiled nature. And it is not a question of physical exercise only, it is also a question of spiritual exercise and enjoyment. It is a question of spiritual values.'[6]

A commonly accepted excuse for apathy about the way the environment is viewed and treated is that our attitudes have been formed by religion (especially Christianity – and communist practice, if that is counted as a religion), and can therefore be rejected as outdated myth. Certainly, most of our irrational fears and unfounded assumptions about the world around us have been dissipated by rationalism and experience. But this does not mean that we live in a sterile bubble. In a highly influential and much reprinted essay on 'the historical roots of our ecologic crisis',[7] US historian Lynn White wrote:

> what people do about their ecology depends on what they think about themselves in relation to things about them. Human ecology is deeply conditioned by beliefs about our nature and destiny – that is by religion. To Western eyes this is very evident in, say, India. It is equally true of ourselves and of our medieval ancestors... [But] especially in its Western form, Christianity is the most anthropocentric religion the world has seen... Christianity, in absolute contrast to ancient paganism and Asia's religions, not only established a dualism of man and nature, but also insisted that it is God's will that man exploit nature for his proper ends.

This dualism between humans and their environment has had a profound effect on our attitudes to the Earth. Through the cultivation of land for agriculture – including the draining of wetlands and consequent changes to the water table by irrigation, deforestation and the removal of prey and pest species by hunting and chemical poisons – we have had a massive impact on natural systems for our own purposes. In the early centuries of agriculture, these effects were

---

[6] Introduction to Dower, J. (1938). *The Case for National Parks in Great Britain.* London: Council for the Preservation of Rural England.

[7] *Science,* **155**: 1203–7, 1967; see also p. 100.

local and limited; their influence was slight. They have increased enormously as technology has advanced. White's conclusion was:

> We are superior to nature, contemptuous of it, willing to use it for our slightest whim... We shall continue to have a worsening ecological crisis until we reject the Christian axiom that nature has no reason for existence but to serve man... Since the roots of our trouble are so largely religious, the remedy must be essentially religious, whether we call it that or not... What we do about our nature depends on our ideas of the man-nature relationship.

White's thesis has been criticized by both historians and theologians, but it remains widely accepted and influential. Max Nicholson's answer to environmental damage was brutal: 'The first step must be plainly to reject and to scrub out the complacent image of Man the Conqueror of Nature, and of Man Licensed by God to conduct himself as the world's worst pest.' In contrast, Texas philosopher Max Oelschlaeger came to accept that White's argument was badly misleading. He wrote:

> For most of my adult life I believed, as most environmentalists do, that religion was the primary cause of ecological crisis. I also believed that various experts had solutions to environmental malaise. I was a true believer. If only people would listen to the ecologists, economists, and others who made claims that they could 'manage planet Earth', we would all be saved. I lost that faith by bits and pieces, especially through the demystification of two ecological problems – climate heating [change] and extinction of species – and by discovering the roots of my prejudice against religion. That bias grew out of my reading of Lynn White's famous essay blaming Judeo-Christianity for environmental crisis.[8]

---

[8] Oelschlaeger, M. (1994). *Caring for Creation*. New Haven, CT: Yale University Press, p. 1.

Who is right: the vastly experienced Nicholson or the relative outsider Oelschlaeger? To be fair to White and to help the debate, he ended his essay with two conclusions, 'Both our present science and our present technology are so tinctured with orthodox Christian arrogance toward nature that no solution for our ecologic crisis can be expected from them alone. Since the roots of our trouble are so largely religious, the remedy must also be essentially religious, whether we call it that or not.'

There is certainly a need for technological solutions to environmental problems, but they are insufficient by themselves. The idea that science is an omnipotent saviour has been dented by too many events in the decades since White wrote – massive oil spills, 'mad cow' disease, volcanic ash disrupting aircraft flights, unexpected earth movements, toxic plastic residues in the ocean. This does not mean that technology has no answers for environmental problems, but many of the suggested 'fixes' (mirrors in space, iron fertilization of the sea, nuclear fusion) are fanciful or not yet attainable. Many religions have explored environmental involvement. Is there any commonality to these endeavours?

Edward Wilson has suggested that there is an instinctive, evolutionarily understandable bond between humans and other living creatures. He calls this 'biophilia'. It implies a human dependence on nature that far exceeds the simple issues of material and physical; it means a human craving for aesthetic, intellectual and even spiritual meaning and satisfaction. René Dubos (p. 42) has made the same point: 'Above and beyond the economic reasons for conservation, there are aesthetic and moral ones that are even more compelling… We are shaped by the earth. The characteristics of the environment in which we develop condition our biological and mental being and the quality of our life.' Biophilia has some likenesses to the physical notion of the 'Anthropic Principle', that physical constants are such that even the smallest deviation would be catastrophic to matter and life as we know it. The astronomer Fred Hoyle was so disconcerted by this that he argued that the universe had the appearance of a 'put-up job', as though 'a super-intellect had monkeyed with physics' to fix all these improbable coincidences.

It may be that the clinical and psychological benefits from even limited amounts of 'exposure' to nature in towns (open spaces, greenery, species richness, etc., which are increasingly being documented) are manifestations of biophilia.[9] If true, they would strengthen the 'something else' expected by Haldane (p. 14). It also returns us to the 'metaphysical questions' Medawar saw as important (p. 12).

Wilson himself sees a convergence between his rationale for conservation and overtly religious considerations. He has written an extended appeal to his hypothetical Baptist Pastor, arguing that their interests come together on caring for creation.[10] Wilson himself was brought up as a Baptist in Alabama, but rejected his childhood faith because of the perceived clash between biological science and the Bible. Notwithstanding, his plea at least raises the question that the Christian conservationist and the secularist share a common concern. The distinguished Kansas historian Donald Worster has set out a middle ground:

> I cannot now recommend that we slip backwards in time and solve the crisis by reading the Bible or Koran again. It is not possible or even desirable to try to go back to a pre-modern religious world-view. We cannot so simply undo what we have become. For this reason I must disagree with Lynn White, who proposed that the world convert to the religious teachings of St. Francis of Assisi, the famous thirteenth-century Italian monk who embraced the plants and animals as his equals and beloved kinsfolk. The idea of making Franciscans of everyone in the world would be an ethnocentric and anachronistic solution to the modern dilemma.

---

[9] Shanahan, D.F., Fuller, R.A., Bush, R., Lin, B.B. and Gaston, K.J. (2015). The health benefits of urban nature: how much do we need? *Bioscience*, **65**: 476–85.

[10] *The Creation*. New York: W.W. Norton, 2006.

Table 10.1 *Yale professor Stephen Kellert has proposed nine values for biophilia (in Kellert and Wilson, p. 59; see 'Further Reading', p. 241). He notes the 'ubiquitous expression of the values', in both his own studies and ones conducted by others. Somewhat intriguingly, they parallel the same factors lying behind the work behind the Earth Charter (p. 211).*

Term	Definition	Value
Utilitarian	Practical and material exploitation of nature	Physical sustenance/ security
Naturalistic	Satisfaction from direct experience and contact with nature	Curiosity, outdoor skills, mental/ physical development
Scientific	Systematic study of structure, function, relationships of the natural world	Knowledge, understanding, observational skills
Aesthetic	Physical appeal and beauty of nature	Inspiration, harmony, peace, security
Symbolic	Use of nature for metaphorical expression, language, expressive thought	Communication, moral development
Humanist	Strong affection, emotional attachment, 'love' for nature companionship	Group bonding, sharing, cooperation,
Moral	Strong affinity, spiritual reverence, ethical concern	Order and meaning in life, kinship and affiliations
Deterministic	Mastery, physical control, dominance of nature	Mechanical skills, physical prowess, ability to subdue
Negative	Fear, aversion, alienation from nature	Security, protection, safety

So what can we do? What is the solution to the environmental crisis brought on by modernity and its materialism? The only deep solution open to us is to begin transcending our fundamentalist world-view by creating a post-materialist view of ourselves and the natural world, a view that summons back some of the lost wisdom of the past but does not depend on a return to old discarded creeds. I mean a view that acknowledges that all scientific description is only an imperfect representation of the cosmos, and acknowledgement that this is the foundation of respect.[11]

Edmund Burke is supposed to have declared almost three centuries ago that 'the only thing necessary for the triumph of evil is for good men to do nothing'. The ball is in our court. The problem is that the health and probably the fate of the planet is also in our court.

FURTHER READING

Holdgate, M. (1996). *From Care to Action.* Gland: IUCN.

Houghton, J. (2013). *In the Eye of the Storm.* Oxford: Lion.

Jenkins, W. and Chapple, C.K. (2011). Religion and environment. *Annual Review of Environment and Resources*, **36**: 441–63.

Jones, N. (2006). *The Plimsoll Sensation.* London: Little, Brown.

Kellert, S.R. and Wilson, E.O. (eds.)(1993). *The Biophilia Hypothesis.* Washington, DC: Island Press.

Medawar, P. (1984). *The Limits of Science.* New York: Harper & Row.

Oreskes, N. and Conway, E.M. (2010). *Merchants of Doubt.* New York: Bloomsbury.

Wilson, E.O. (1984). *Biophilia.* Cambridge, MA: Harvard University Press.

(2006). *The Creation.* New York: W.W. Norton.

[11] Worster, D. (1993). *The Wealth of Nations.* New York: Oxford University Press, p. 218.

11 From Scavenging to Supermarkets

Before the Roman came to Rye or out to Severn strode,
The rolling English drunkard made the rolling English road.
A reeling road, a rolling road, that rambles round the shire…

G.K. Chesterton

We noted at the beginning of this book that our relationship and therefore our attitudes to the environment were complicated. We have 'rolled drunkenly' in getting to where we are. We have also seen that our environment is much more than a convenient envelope around us. In his penetrating analysis of global geopolitics, Tim Marshall has pointed out the limitations imposed upon us:

> The land on which we live has always shaped us. It has shaped the power, politics and social development of the peoples that now inhabit nearly every part of the earth. Technology may seem to overcome the distances between us in both mental and physical space, but it is easy to forget that the land where we live, work and raise our children is hugely important, and the choices of those who lead the seven billion inhabitants of this planet will to some degree always be shaped by the rivers, mountains, deserts, lakes and seas that constrain us all – as they always have.[1]

Marshall was concerned almost exclusively with the physical structure of the world, but his point extends far beyond mere topography. If we are honest, recognizing the influence of our environment should humble and provoke awe in us, even for the avowedly non-religious. Max Nicholson, a militant humanist, writes of growing up,

---

[1] Marshall, T. (2015). *Prisoners of Geography*. London: Elliott & Thompson.

reading the *Natural History of Selborne*, and finding that 'wherever I went… signs of the dynamic and dramatic workings of nature were an unfailing counterpoise to taking it for granted, or looking on it as a picture galley or museum'.[2]

Julian Huxley described his stay on a small Welsh island:

> Here I felt, perhaps even more than in Africa, the power and inde-
> pendence of nature – nature that helps things make themselves, as
> Charles Kingsley wrote in the *Water Babies*. The swarms of puffins
> flying down from the cliffs and resting on the sea, the screeching of
> guillemots, the great black-backed gulls screaming and devouring
> the plump young shearwaters as they stumbled to the cliff-edge
> before attempting their first flight, yet (if they survived the pred-
> atory gull's attack) immediately at home and knowing what to do
> when they reached the water; the occasional gannets soaring on
> their wide white wings: all these manifestations of the vast inter-
> related web of life never ceased to provoke my interest and wonder.[3]

Composer Peter Maxwell Davies went to Antarctica in 1997–8 – a trip that provided the inspiration for his *Antarctic Symphony* – and noted in his diary 'that Antarctica reminds me more than anything of the hidden artwork in medieval cathedrals created by sculptors to the greater glory of God. One is unaccustomedly hypersensitive there to the act of Creation. Elsewhere on earth, man is the most successful animal; in Antarctica, wonderfully, he has only a precarious toehold.' Towards the end of his life, Goethe declared that a sense of astonish-ment or wonder was an end in itself; we should not seek anything beyond or behind this experience, but simply enjoy it for what it is. Einstein would have agreed. He called 'rapturous amazement' a key to living.

Arthur Tansley believed that there is 'a real and deep aesthetic satisfaction brought by the constant contemplation of the infinite

---

[2] Nicholson, E.M. (1970). *The Environmental Revolution*. London: Hodder & Stoughton, p. 18.

[3] Huxley, J.S. (1970). *Memories*. London: Allen & Unwin, p. 218.

beauty of Nature, ranging from the galaxies of astronomical space through the marvellous diversity of scenery and of the living things inhabiting the earth's surface, to the fascinating world revealed by the microscope'. He was clear that there exists 'an aesthetic aspect of science, which opens the mind to a far deeper and wider realization of the wonder and beauty of the universe [and] that this is not a trivial claim, though it may seem to some old-fashioned'.[4]

In her last book, Rachel Carson, discomforter of the US chemical establishment, wrote of taking her two-year-old nephew to the New England shore on a stormy day:

> Together we laughed for pure joy – he a baby meeting for the first time the wild tumult of Oceanus; I with the salt of half a lifetime of sea in me. But I think we felt the same spine-tingling response to the vast, roaring ocean and the wild night around us... It is our misfortune that for most of us that clear-eyed vision, that true instinct for what is beautiful and awe-inspiring, is dimmed and even lost before we reach adulthood... Is the exploration of the natural world just a pleasant way to pass the golden hours of childhood or is there something deeper? I am sure there is something much deeper, something lasting and significant...[5]

Even Richard Dawkins, the high prophet of atheism, does not deny that there are 'impulses to awe, reverence and wonder'. He regards them as precisely the same impulses 'that led the poet William Blake to Christian mysticism, Keats to Arcadian myth and Yeats to Fenians and Fairies', but he believes it is 'the very same spirit that moves great scientists'. For the scientist, 'our interpretation is different but what excites us is the same. The scientist has the same wonder, the same sense of the profound as the mystic, but with an additional impulse: let's find out what we can about it'[6] (Figure 11.1).

[4] Tansley, A.G. (1942). The values of science to humanity. *Nature*, **150**: 104–10.

[5] Carson, R. (1965). *The Sense of Wonder*. New York: Harper and Row, pp. 15, 54, 100.

[6] Dawkins, R. (1998). *Unweaving the Rainbow: Science, Delusion and the Appetite for Wonder*. Boston, MA: Houghton Mifflin Harcourt.

FIGURE 11.1A–C Attitudes of awe. A. Loch Coruisk in Skye, looking towards the over-sensationalized Cuillin Ridge, from William Daniell's *Voyage Round Great Britain* (1814–25). B. Watercolour painting of Yosemite Park by Doris Jung-Rosu. C. Ocean waves – always a source of awe. Image: hitforsa.

FIGURE 11.1A–C (CONTINUED)

We have travelled far in the five million years or so since we came down from the trees. It has been a circuitous route with many cul-de-sacs, many of them long, alluring and sometimes fatal to their explorers. We began as scavengers on the plains of Africa; we now occupy almost all the habitable places on Earth. We may or may not have experimented with living in the sea (p. 24). Our original migrations from our African home were probably tentative. When we reached what seemed to be a green and pleasant land, our settlements often turned out to be at the mercy of a changing climate or a fragile environment prone to erosion. We have had to learn what crops to plant and how to nurture them, and what animals we could domesticate and use. There must have been many failures. Sometimes whole communities perished, as seems to have happened to the Norse in Greenland. Sometimes only a few survived, as in the case of the Black Death. The only constant factor has been the attempt to protect ourselves by achieving independence from the environment.

Unfortunately, and perhaps dangerously, this environmental eman-cipation has been achieved at what has been called the 'extinction of experience'. In the so-called civilized world, fewer of us interact with the natural world than we used to. Never mind the natural capital we are losing, we seem to have put ourselves into a potential downward spiral of our own well-being. Perhaps we should learn from the need for children to be exposed to various diseases at a young age to sur-vive better in later life.

It is easy to be overpessimistic. We ought to have learnt enough from our past to see something of the road ahead of us. The tragedy is that we repeatedly fail to make use of this knowledge. We still eat or drink unwisely. We may have discovered how to treat our own sewage and even use it as a fertilizer, but we still pollute widely with dangerous chemicals, plastics and noxious gases. We may be aware that our decisions will cause problems for others, such as increasing greenhouse gases by our own travel methods or demands, or objecting to developments (power lines, windmills, waste processing plants) for our convenience or comfort. This has not prevented us bulldozing our chosen path and thereby too often causing upsets to others, whether in our own neighbourhood or in a far country, or – perhaps least forgivably – our children. Does growing food have pri-ority over preserving the habitat of pandas or dragonflies? We have moved from an existence where we could not choose our fate to one where we can. We now have the information that makes for well-being, although we too often fail to use it properly. We do not trust our neighbours. A survey of the values claimed by Britons found that 74 per cent claimed greater importance for compassion in themselves rather than selfishness, but more (77 per cent) believed that their fellows held selfish values.[7] What are we to conclude from this? The sad likelihood is that we will not behave altruistically if we see our sacrifice negated by others.

[7] Common Cause Foundation (2016). *Perceptions Matter: the Common Cause UK Values Survey.* London: Common Cause Foundation.

Does all this all sound like pious utopianism or a grieving Luddite's lament? I would like to think we might read it as an invitation for moral awareness. As part of nature, we are subject to the chances and vicissitudes of a dynamic world, but we are also apart from nature, able to stand apart and take account of not only our own interests but also those of our family and community and, indeed, of the world itself. Despite having lost our shaggy body hair, we too often behave like the hairy apes we left behind several million years ago. We are animals; the challenge is to behave also as conscious and moral agents.

Is our behaviour irretrievably hard-wired or can we go beyond to reason? As the Spanish philosopher José Ortega y Gasset (1883–1955) put it, 'Scientific truth is characterized by its precision and the certainty of its predictions. But science achieves these admirable qualities at the cost of remaining on the level of secondary concerns, leaving ultimate and decisive questions untouched.'[8] The great and good of Victorian times not uncommonly embellished their buildings with texts or pious inscriptions. The University Museum in Oxford, founded in 1860 as a centre for science in Oxford, and the Natural History Museum in London are both built in a grand ecclesiastical style, with all sorts of religious emblems (Figure 11.2).

The museum buildings are products of a high Victorian age and its beliefs, but the faith they represent did not die with Queen Victoria. The Cavendish Laboratory in Cambridge has been a centre of scientific discovery from its earliest years, with twenty-nine Nobel Laureates awarded for research carried out within it. Its first Director, James Clerk Maxwell, who formulated the theory of electromagnetic radiation, decreed that a quotation from one of the biblical psalms be carved on the gate of the newly founded Laboratory, 'Great are the works of the Lord, studied by all who delight in them.'[9]

[8] The sportive origin of the state. In: *Toward a Philosophy of History.* New York: W.W. Norton, 1941.
[9] Psalm 111, verse 2.

FIGURE 11.2 Manufactured attitudes. The Oxford University Museum of Natural History was built in 1855–60 as a 'temple of science'. The stone frieze around the main entrance depicts (in an unfinished way – the builders ran out of money) the works of creation, surmounted by an angel holding an open book in one hand and three living cells in the other, 'to signify the intention of the founders of the museum whose desire it was to bring future generations of men to study the open book of nature and the mysteries of life under the guidance of a higher power which alone could enable them to read the pages of that book with right understanding'. The Oxford Museum pre-dated the Natural History Museum in South Kensington by twenty years, but the latter was also designed (according to *The Times*) as 'a Temple of Nature, showing the Beauty of Holiness'.
Photo: Andrew Berry.

When they worked out the structure of DNA in 1953, Francis Crick and Jim Watson rushed out through this gate to the pub across the street 'to tell everyone within hearing that we had found the secret of life', although it is unlikely that they took in the words above them. Twenty years later, the Cavendish Laboratory moved to a new site on the edge of Cambridge. Sir Brian Pippard, the Cavendish Professor at the time, has written, 'Shortly after the move to the new

FIGURE 11.3A The Entrance to the old Cavendish Laboratory (Physics Department) in central Cambridge. Photo: Noah Tate.

buildings, a devout research student suggested to me that the same text should be displayed, in English, at the entrance. I undertook to put the proposal to the Policy Committee, confident that they would veto it; to my surprise, however, they heartily agreed to the idea.' The text is on the door of the 'new' Cavendish Laboratory (Figure 11.3).[10]

The Cavendish Laboratory and its exuberant offspring, the Laboratory of Molecular Biology, can be regarded as continuing a tradition of an experimental science which began with Aristotle (p. 64).

[10] Berry, R.J. (2008). The research scientist's psalm. *Science & Christian Belief*, **20**: 147–61.

FIGURE 11.3B When the original wooden doors are closed, the words of Psalm 111:2 ('Great are the works of the Lord; studied by all who delight in them') are found. The Psalm was added by decree of the first Cavendish Professor, James Clerk Maxwell.
Photo: R.S. White.

Along the road was Franciscan friar Roger Bacon (1219–92), who studied in Paris at the same time as Thomas Aquinas was working on integrating Aristotle's ideas with Christian teaching. Bacon's concern was to distinguish natural philosophy (which proposes theories) from 'experimental science', which investigates them. He recognized two sorts of experience: an internal knowledge of things spiritual which 'comes from grace' and one which is 'gained through our external senses'.

A similar idea surfaced three centuries later from another Bacon – Francis – who laboured to replace traditional learning with a new system based on empirical and inductive principles. Darwin prefaced *On the Origin of Species* with a quotation from this second Bacon: 'Let no one think that they can search too far or be too well studied in the book of God's words [the Bible] or in the book of God's works [nature]; rather let everyone endeavour a endless progress or proficiency in both.' It is trite to record that Darwin was not a theologian. He wrote to his friend Joseph Hooker, 'My theology is a simple muddle [but] I cannot look at the universe as a result of blind chance.' Richard Dawkins has argued that 'although atheism might have been tenable before Darwin, Darwin made it possible to be an intellectually fulfilled atheist'.[11] While this may be true, Darwin himself wrote, 'It seems to me absurd to doubt that a man may be an ardent Theist and an evolutionist... I have never been an atheist in the sense of denying the existence of a God.'[12] It is commonly believed that he went through a crisis of faith when his daughter Annie died at the age of ten, but in his *Autobiography*, written many years later, he insisted that he always believed in God.

Can we be more specific about the attitudes which led to the motifs on the London and Oxford Museums and the Cavendish Laboratory? It has been said that the curiosity which drives science has 'throughout history swum in the slipstream of an ultimate metaphysical curiosity rooted in the human need to make sense of the world as a whole'.[13] This has involved at least four ideas:

- the existence of a beneficent and rational agency;
- that this agency is not identifiable with anything within the universe, but gives the whole a law-like character;
- that truth is not the prerogative of any one civilization; and
- that this truth involves the need for testing to validate it.

Darwin, Huxley, Maxwell Davies, Nicholson and Carson have written of their personal reactions of awe. Their beliefs strengthen the likelihood that a description of the environment in terms only of biology, geology, atmosphere and history is incomplete. This

---

[11] Dawkins, R. (1986). *The Blind Watchmaker*. London: Longman, p. 6.
[12] Letter to John Fordyce, 7 May 1879.
[13] Wagner, R. and Briggs, A. (2016). *The Penultimate Curiosity*. Oxford: Oxford University Press, p. 411.

does not prove anything, but it would be wrong to ignore them. Cumulatively, it highlights the danger of reductionism in our environmental attitudes. The commentator and scientist Steve Jones wrote in his Reith Lectures:

> It is the essence of all scientific theories that they cannot resolve everything. Science cannot answer the questions that philosophers – or children – ask: why are we here, what is the point of being alive, how ought we to behave? Genetics has almost nothing to say about what makes us more than machines driven by biology, about what makes us human. These questions may be interesting, but scientists are no more qualified to comment on them than is anyone else. In its early days, human genetics suffered greatly from its high opinion of itself. It failed to understand its own limits.[14]

William James made the same point: 'Science can tell us what exists; but to compare the *worths*, both of what exists and what does not exist, we must consult not science, but what Pascal calls our heart.'[15]

The German astronomer Johannes Kepler (1571–1630) had no doubts about this. As a young man he wrote to a friend, 'I wanted to become a theologian; for a long time I was unhappy. Now, behold, God is praised by my work, even in astronomy.' For him the practice of science was 'thinking God's thoughts after him'. His prayer at the end of *Harmony of the World* (1619) was:

> If I have been enticed into brashness by the wonderful beauty of thy works, or if I have loved my own glory among men, while advancing in work destined for thy glory, gently and mercifully pardon me: and finally, deign graciously to cause that these demonstrations may lead to thy glory and to the salvation of souls, and nowhere be an obstacle to that. Amen.

---

[14] Jones, J.S. (1993). *The Language of the Genes*. London: HarperCollins, p. xi.
[15] James, W. (1897). *The Will to Believe and Other Essays*. New York & London: Longmans Green, p. 27.

Kepler's world was very different from ours. What about J.B.S. Haldane's conclusion: 'I have no doubt that in reality the future will be vastly more surprising than anything I can imagine. Now my own suspicion is that the Universe is not only queerer than we suppose, but queerer than we *can* suppose.'[16] Kepler belongs to a different era and mindset from us, but it is worth recognizing that his acknowledgement of what Dawkins calls awe, reverence and wonder is probably the same as the queerness suspected by the Marxist Haldane.

Lynn White castigated his fellows for apathy about the environment as a result of their religious ideas, but, as we have seen, he nevertheless concluded, 'the remedy must also be essentially religious, whether we call it that or not'. He was not writing about the dogmatic fatalism common in Islam or the equally dogmatic belief of some evangelical Christians that they can neglect the Earth in order to work only towards a future Rapture when they will caught up to heaven, rescued from a cursed Earth. These are extreme views, but from earliest human times, human attitudes to the environment have joined pragmatism with religion of some sort. Many have drowned in the myth of dogmatic certainty; perhaps as many again have tortured themselves with uncertainty. Peter Medawar had no doubt about the existence of 'metaphysical questions' (p. 12). Formal religion is an irrelevance for most of us, but that does not stop us asking metaphysical questions about the slings and arrows of outrageous fortune or, indeed, the meaning of life itself.

The one generality that seems to emerge from our incident-laden but successful journey is the reality of something that can be described as religion in the widest understanding of that term. Can we be more specific? Is there anything that connects the widely recognized awe of the natural world with the queerness of the world as perceived by Haldane and the possibility of 'metaphysical questions' foreseen by Medawar, never mind the need for religion of some sort that White identified as necessary for modifying our environmental

[16] *Possible Worlds*. London: Chatto & Windus, p. 286.

attitudes? Environmentalists often claim rather glibly that everything is connected. As Jonathan Swift put it three hundred years ago
in 'On Poetry: a Rhapsody', 'So naturalists observe, a flea Has smaller
fleas that on him prey; And these have smaller still to bite 'em, And
so proceed *ad infinitum*.' This is effectively a description of the web
of life concept developed from actual observations by Alexander
von Humboldt and Wolfgang Goethe, adopted in different ways by
followers like Ernst Haeckel, John Muir, David Thoreau and Wendell
Berry. Humboldt wrote, 'In this great chain of causes and effects, no
single fact can be considered in isolation.'[17] As we have seen, Darwin
was a great admirer of Humboldt, but with the 'additional impulse'
which Dawkins identifies as characterizing science, his contribution
was to identify natural selection as the major driver of evolutionary
change. Humboldt foresaw the damage humans were causing, but his
vision and achievements have been largely forgotten.

How does religion (used as above in a wide sense) relate to
science? Religions differ in their understanding of their relationship with science, but there is general agreement that scientific
explanations should not be confused with religious ones. Scientists
are concerned with mechanisms, with how things work. When the
French astronomer, the Marquis de Laplace was asked by Napoleon
about the place of God in his science, he replied 'Sire, I have no need
for that hypothesis.' He was not denying the existence of God (he
was a devout Roman Catholic), but testifying that there was no place
for a sort of treadmill operator in his understanding of the universe.
He was using the truth realized by Aristotle, that an event can have
more than one cause (p. 231). The Danish physicist Niels Bohr introduced the term 'complementarity' to describe the contradiction that
light and electrons can apparently behave contradictorily as either
particles or waves. Other scientists have seen complementarity as a
valuable concept in a range of contexts. Robert Oppenheimer applied

---

[17] *Essay on the Geography of Plants*, 1807. Humboldt is also said to have anticipated
Gaia, which is a major theory of interconnectedness (p. 174), describing the earth
as 'a natural whole animated and moved by inward forces'.

it to mechanistic versus organic analyses of life processes; Charles Coulson to problems of mind versus brain, free will versus determinism, theology versus mechanism; William Pollard to human freedom versus divine providence. The brain scientist Donald MacKay has explored its use in Christian debates about science and religion. He wrote:

> The God in whom the Bible invites belief is no 'cosmic mechanic'. Rather is he the Cosmic Artist, the creative Upholder, without whose constant activity there would not be even chaos, but just nothing. What we call physical laws are expressions of the regularity that we find in the patterns of created events that we study as the physical world. Physically, they express the nature of the entities 'held in being' in the pattern. Theologically, they express the stability of the great Artist's creative will. Explanations in terms of scientific laws and in terms of divine activity are thus not rival answers to the same question; yet they are not talking about different things. They are (or at any rate purport to be) complementary accounts of different aspects of the same happening, which in its full nature cannot be adequately described by either alone.[18]

FURTHER READING

Bellah, R.N. (2011). *Religion in Human Evolution.* Cambridge, MA: Harvard University Press.

Berry, R.J. (2003). *God's Book of Works.* London: T&T Clark.

Glacken, C.J. (1967). *Traces on the Rhodian Shore.* Berkeley, CA: University of California.

Moo, J. and White, R. (2013). *Hope in an Age of Despair.* Nottingham: IVP.

Wagner, R. and Briggs, A. (2016). *The Penultimate Curiosity.* Oxford: Oxford University Press.

[18] MacKay, D.M. (1960). *Science and Christian Faith Today.* London: Falcon.

Index